Matthew Watkinson qualified as a veterinary surgeon in 2001 and then spent the next eight years realising that he was specifically fighting natural selection and thus making welfare problems worse, not better. This led him to examine the relationship between compassion and suffering and eventually to the emotional corruption of Charles Darwin's theory of natural selection that has ultimately inspired this book.

THE DESTINY OF SPECIES

by Means of Natural Selection,

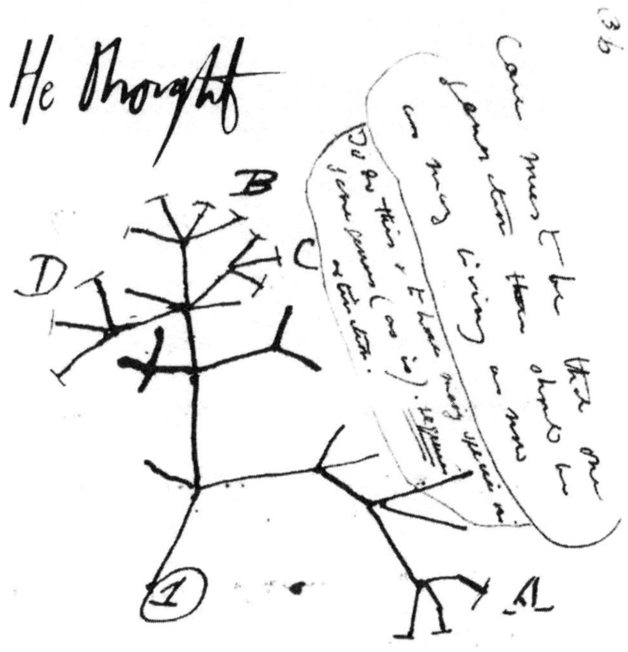

or the

Elimination of Unfavoured Races in the Struggle for Life.

Matthew Watkinson

Matador
5 Weir Road
Kibworth Beauchamp
Leicester LE8 0LQ, UK
Tel: 0116 279 2299
Fax: 0116 279 2277
Email: books@troubador.co.uk
Web: www.troubador.co.uk/matador

ISBN 978 1848763 067

British Library Cataloguing in Publication Data.
A catalogue record for this book is available from the British Library.

Front cover: 'Arktisk Fauna'. *Bibliographisches Institut,* Leipzig (from the
Nordisk Familjebok (1915), vol.21, *Till art. Polarländer*).

Printedf in the UK by TJ International, Padstow, Cornwall

Matador is an imprint of Troubador Publishing Ltd

To my friends:

Please don't take any of this personally.

CONTENTS

Chapter 1

INTRODUCTION

When on board Planet Earth, as a naturalist, I have been much struck by humanities gushing and often paradoxical sentimentality. It seems almost ubiquitous and one example that really stands out occurred during my final year of veterinary training. I have forgotten the name of the dog of course, and indeed a lot of other ancillary details, but I do know that its front leg had been amputated to remove an aggressive bone tumour, and I do know that I will never forget its screams. It wasn't even a young dog. It was an old dog with cancer and yet, despite being within touching distance of the end, it was lying in a soulless hospital kennel screaming in agony and recoiling in horror when anybody approached. It was horrible and no matter how hard I tried I couldn't justify its suffering.

I must admit, I couldn't justify my own anger properly either. I was quite sure the dog's suffering was entirely based on the emotional needs of its owners, and that it couldn't have suffered if it had been euthanased, but the full implications were beyond me. What if the dog had been younger for instance, would that have made the suffering acceptable? And if it had, what does 'younger' actually mean? Is it less than 8 years old? Or 9? Or 8½? And if it had been young enough to make the procedure acceptable, does a dog know whether the pain will stop, even if I do? Indeed, can I actually guarantee that the pain *will* stop (and that long term gain will *definitely* follow short term suffering), or is there a significant probability that the treatment will fail and render the whole attempt worse than just killing the animal in the first place? In fact, what is the problem with killing animals in the first place anyway, regardless of age?

I had no idea at the time, and when I was inevitably summoned by the clinical elite for a 'chat' I was unable to justify my anger properly.

I wouldn't have the same problem anymore. In fact, I would relish the opportunity for a 'chat' these days because it's now quite clear to me that emotion severely corrupts perspective. It twists judgement and warps rational thought and the journey I've been on to understand this case, and a whole host of others, has taken me further and further from compassionate sentimentality and right back to the system that has worked for more than three and a half thousand million years. It has taken me right back to the totally unsentimental and utterly ruthless world of natural selection. In particular, I have learnt to see organic beings through their own eyes and even though that simple shift in perspective has totally undermined my veterinary career, it has also made me realise that humans don't define reality, and that death is a vital part of life.

It took me a while to put two and two together of course, but the journey is defined by its destination and it's now quite clear that our relationship with nature has been severely corrupted by paradoxical emotion and unjustifiable self-reverence. Despite the evidence, we continue to believe that we know best and that's why I have written this book. *On The Destiny of Species* is my assessment of our subjective beliefs and I will use logic, common sense and Charles Darwin's *On The Origin of Species* throughout to challenge everybody who claims to 'love' domestic animals.

I will go much further than that though, because despite making a dreadful mess of the domestic species many now claim that we're here to manage *all* animals. In fact, many now claim we're here to manage the entire planet and in this book I will also use logic, common sense and *On The Origin of Species* to challenge that extreme narcissism too. I will conclusively destroy conservation myths about the fragility of Life, the sanctity of species, the iniquity of pests, the reality of sustainability, the relevance of appearance, the importance of scarcity, the majesty of conservation, and indeed, the empirical basis for almost everything the conservation community has ever said. Whatever they may think, managing nature (conservation) is not observing nature (natural history) and observing nature is the key to the truth.

Before we begin, I do need to make a few things clear.

Firstly, I'm not, under any circumstances, advocating a specific ethical philosophy. In fact, I believe that ethical diversity is as natural as biological diversity, but either way, I'm *not* preaching. If you want to take 8000 long haul flights a year, or buy fish that have been deliberately gutted alive, or recycle a few newspapers, or drive a petrol-guzzling urban tank, or mourn dead animals that couldn't care less, or continue to believe that humans are the centre of the Universe, that's your decision. You can say whatever you like and do whatever you like but, and this is the critical part, if it isn't objective, it's not natural history. If your actions and opinions have been forged in a cauldron of self-indulgent emotion, you can preach from the pulpit of your own opinion as much as you like, but you can't legitimately claim to walk in the footsteps of impartial naturalists like Charles Darwin, and you definitely can't legitimately claim to understand the brutal reality of life on Earth.

Secondly, I would like to mention post-rationalisation, because challenging pre-conceived opinions always upsets people and because upset people who have had their opinions challenged often post-rationalise. For those who don't know, post-rationalisation is the brain's ability to justify old beliefs in the light of new evidence. It's a psychological defence mechanism and it will, because it always does, lead to much wounded trumpeting. Typically this will involve blind faith and wild hope, but whatever the technique I will say this: almost everything I have to say is supported by *The Origin of Species,* which means cynics will have to disagree with Charles

Darwin to disagree with me.

Talking of Charles Darwin, I should also point out that the sixth edition of *On The Origin of Species by Means of Natural Selection or the Preservation of Favoured Races in the Struggle for Life** will feature throughout and that all quotes that aren't specifically introduced will be labelled with a superscript [Darwin]. For example:

> 'Although I am fully convinced of the truth of the views given in this volume under the form of an abstract, I by no means expect to convince experienced naturalists whose minds are stocked with a multitude of facts all viewed, during a long course of years, from a point of view directly opposite to mine…I look with confidence to the future,—to young and rising naturalists, who will be able to view both sides of the question with impartiality.'[Darwin]

Thirdly, I would like to warn those of a squeamish disposition. Life on Earth is *not* a harmonious world of gum-drop smiles and pretty flowers; it's an unsentimental death fight. It's a battle for life and though this may upset the faint of heart, and the pampered of constitution, I will make no apologies for this. This is about objective natural history, not privileged sensitivities, and you have been warned. If you can't cope with reality, *don't* read this book.

Penultimately, I would like to alert the creationist community. I don't want to waste anybody's time and I can categorically state that this isn't a book for people who need the Universe to love them in a special way. It hasn't been written by biased scholars with lots of superstition and no evidence and thus it won't provide any spiritual satisfaction whatsoever. Of course, if there's even a faint glimmer of rational

---

* All editions of the *The Origin of Species,* and indeed everything else Charles Darwin ever wrote, are available to check online at www.darwin-online.org.uk.

enquiry please read on, but this book is primarily aimed at people who believe in the miracle of reality, rather than the reality of miracle, and if you can't accept evolution, I will salute you as a member of Life's rich variety and urge you to stick with the supernatural fantasy books you clearly know and love.

Finally, I would like you to reserve judgement. I have already made claims bold enough to generate a reaction, but please don't dismiss them yet. All you need to do at this point is relax all pre-conceived beliefs and approach the rest of this book with an open mind. You can be a sceptic if you wish, but please don't dismiss the case before it's been heard. In fact, you should be a sceptic. I don't want you to be an unyielding cynic of course, but I would much rather you checked my references, and read *The Origin of Species*, and studied the fossil record and went outside to see nature for what it really is because passive acceptance of other people's opinions is exactly why I have had to write this book in the first place. It's the reason Life is currently regarded as a dependent infant, rather than a magnificent warrior, and so you should be a sceptic. It's the only way to separate faith from knowledge and that's critically important, because I don't want you to believe; I want you to know.

Chapter 2

ORIGIN

I'm not a big fan of grey. A lot of people are but I'm not. It seems to complicate everything and I prefer to see the world in black and white. I think it's called bivalence in classical logic but either way, it's always worth remembering that conflicting perspectives can often be divided into two opposite and mutually exclusive possibilities. For instance, when we consider the Earth's cosmic significance there are really just two sides to the debate: one, the Earth *is* the centre of the Solar System, or two, it's not. Similarly, when we consider our cosmic significance there are also just two sides to the debate: one, we *are* the centre of the Universe, or two, we're not. And the origin of species is no different. If we're going to consider the destiny of species we must first understand the origin of species and luckily for us there are just two possible sides to that debate as well: one, species *have* changed over time, or two, they haven't.

Which is it though? Planet Earth is full of incredible adaptations and wondrous forms but where do they come from? What generated the structure of Umbrella-Mouthed Gulper Eels for instance, with their vast mouths so admirably adapted to catching their vast prey? Was it the gradual accumulation of slight and beneficial variations, or the instant provision of perfect and rigidly defined similarities? And what about Giant Sequoia trees? Has their phenomenal potential evolved over time, or endured despite time?

What is the true origin of species?

Creationists will tell you they have all been designed and manufactured by a benevolent creator of one sort or another. Based on very old books they will argue that descent with modification is demented nonsense and that the 'preservation of favourable individual differences and variations'[Darwin] is blasphemous theoretical porridge. They will be utterly convinced that the spontaneous generation of rigid, immutable species by a supernatural creator who may, or may not, have a big beard is far more sensible.

But is it?

No. It's not. It's not sensible at all. The hardcore religious may be convinced, but in reality immutable creations that can't change at all, or micromutable creations that can't change very much[*], could only survive if the conditions of life were constant, or mildly inconsistent, and they're not. They're changing all the time, on many different levels.

At the cosmic level complex gravitational interactions between the Sun, the Moon and the planets (particularly Jupiter and Saturn) cause orbital deviations that change the Earth's relationship with the Sun. These Milankovitch cycles[†] involve changes in the shape of the Earth's orbit and changes in the tilt and orientation of the Earth's axis that ultimately lead to solar modifications and cyclical climate change; from icehouse to greenhouse and back again, with everything in between.

At the planetary level vast tectonic plates are drifting around the planet in a slow (they typically move 50-100 mm/year) geographical ballet that is constantly changing the environmental conditions of life. They're forming mountains and changing weather dynamics; or moving continents and changing ocean currents; or moving each other and causing earthquakes; or moving apart and causing volcanic eruptions.

At the atomic level molecular gradients are constantly changing global chemistry and altering the conditions of life.

---

[*] Micromutable isn't a real word, and microevolution isn't a real limit. It's a theological concession allowing species to change a little bit, but under no circumstances are they allowed to become different species (macroevolution). That would be a step too far (even though the process is exactly the same).

[†] Milankovitch cycles are named after Milutin Milanković, the Serbian mathematician who first described them.

They're moving solutes and changing the chemistry of the oceans; or dissolving gases and changing the chemistry of the atmosphere; or weathering rocks and changing the geography of the planet, and so it goes on. For a myriad of complex inorganic reasons the conditions of life are constantly changing and there are plenty of organic reasons as well, because Life interacts with itself in all sorts of complex ways. For instance, as soon as one species starts eating another species a dynamic relationship is born. Increased numbers of the resource species mean increased numbers of the consumer species, but increased numbers of the consumer species eventually leads to decreased numbers of the resource species, and that eventually leads to decreased numbers of the consumer species, which then allows increased numbers of the resource species and so it goes on. And if a different consumer species is eating the original consumer species, and the original consumer species is also eating another resource species, which is also being eaten by another consumer species, which is parasitised by another species, which is transiently hosted by another species etc. etc., you have a complex food chain that can change in an instant, because of time based dynamics that are highly sensitive to initial conditions[*].

Then there are Life's effects on its own raw materials, like oxygen and carbon dioxide, and its interactions with its own waste products, like oxygen and carbon dioxide, and when you put everything together the whole thing is a big complex mess of interweaved variables that may, or may not, affect some, or all, of everything or something. In the words of Charles Darwin:

---

[*] Time based dynamics that are highly sensitive to initial conditions are described by chaos theory. Chaos theory isn't as random as the name implies though. It actually refers to very tiny variations in the initial conditions that lead to very large effects as time goes by. For instance, if you place a football at the top of a mountain, very tiny differences in the initial resting position can lead to very large differences in the final resting position. A millimetre either way can mean the football rolls down one side of the mountain or the other, and so the time based result is very sensitive to the initial conditions.

'We must never forget how infinitely complex and close-fitting are the mutual relations of all organic beings to each other and to their physical conditions of life.'

They're staggeringly complex and the simple fact is that change is inevitable. The organic and inorganic conditions of life are dynamic, not fixed. They're changing all the time and organic beings must be able to change with them, because to survive through change they must be able to change through time and thus they cannot be lineal descendants of some same and generally similar species, they must be lineal descendants of some other and generally extinct species. They must have evolved.

And they have. Life has come a long way and species have changed a great deal. The evidence bears great testimony to this fact and the result is the great Tree of Life:

'The affinities of all the beings of the same class have sometimes been represented by a great tree. I believe this simile largely speaks the truth. The green and budding twigs may represent existing species; and those produced during former years may represent the long succession of extinct species.'[Darwin]

We even know quite a lot about the great roots of the great Tree of Life. Darwin didn't, but we do. He was well aware that all living things have much in common, and he definitely suspected that all animals and plants are descended from some one prototype, but he couldn't be absolutely sure and his first Tree of Life was based on logic rather than science. He didn't have any evidence, but we do. We can't identify the precise origin of the precise individual of the precise form yet, but we have discovered ancient unicellular microfossils and we do know that the mummy and daddy of all modern Life (more commonly known as the last universal common ancestor) was a single-celled micro-organism with great flexibility and great potential. It began its journey at least 3.5 *billion* years ago and it has been evolving ever since.

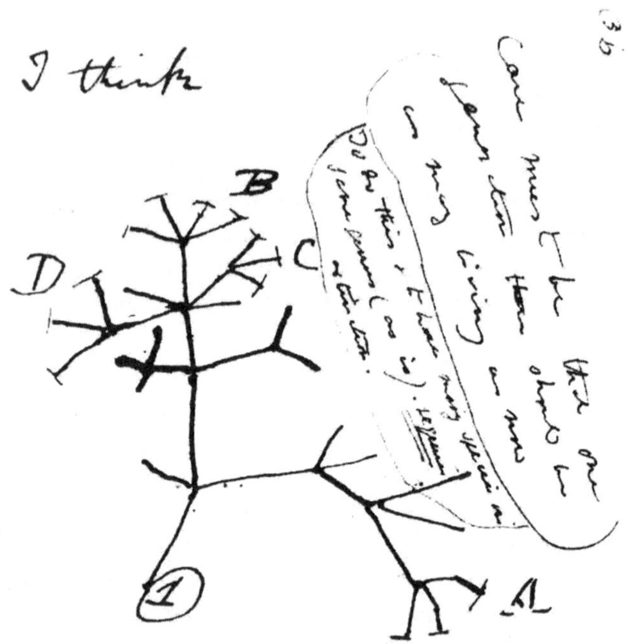

Charles Darwin's first Tree of Life.[*]

To be fair, Darwin's Tree of Life is more of an impenetrable Bramble Thicket of Life. The green and budding twigs may still be green and budding, but they probably merge and split in a knotted mess of green twigs and common purposes. In reality, biological systems are always more complex than anyone can imagine, but the simile still has value and a bramble thicket still has roots. Life may not be the tidy branching ascendency that Darwin first imagined, but all modern life did

---

[*] Charles Darwin: *Notebook B* (1837).

start with one common primordial form, and it is still evolving today.

To be honest, it's not even a debate anymore. Well, not amongst people who value evidence anyway. For those who trust fantasy resistance remains high, but they can believe whatever they like, because they don't define reality. Luckily for life on Earth, and that includes *Homo sapiens*, reality defines itself and for people who respect evidence the mechanics of evolution are quite clear:

> 'As many more individuals of each species are born than can possibly survive; and as, consequently, there is a frequently recurrent struggle for existence, it follows that any being, if it vary however slightly in any manner profitable to itself, under the complex and sometimes varying conditions of life, will have a better chance of surviving, and thus be naturally selected.'
> Darwin

Which basically means that variation, combined with pressure to survive, makes the preservation of favourable individual differences and variations logically inevitable. It means that favoured variations are more likely to survive to reproduce and thus more likely to enter future generations. And that unfavoured variations are less likely to reproduce and thus less likely to enter future generations. For instance, if slow Cape Fur Seal pups are killed by Great White Sharks, they won't survive to breed and variation in ability will lead to biased reproduction and selection for shark evasion. And if herbivores are eliminated if they can't reach elevated foliage, variation in ability will lead to biased reproduction and selection for Giraffe. And if hominids are eliminated if they can't use fire, variation in ability will lead to biased reproduction and selection for pyro-competency.

Conversely, if the conditions of life change so will the results. If slow Cape Fur Seal pups *aren't* killed by Great White

Sharks they *will* survive to breed and variation in ability will lead to biased reproduction that selects for energy efficiency rather than shark evasion. And if herbivores *aren't* eliminated if they can't reach elevated forage, variation in ability will lead to unbiased reproduction and indifference to Giraffe. And if hominids are eliminated if they *can* use fire, variation in ability will lead to biased reproduction and selection for pyro-incompetency. There is no end-point. There is no target. Natural selection will not produce absolute perfection. It's not even about the relentless pursuit of complexity. It's simply about surviving:

> '…for natural selection, or the survival of the fittest, does not necessarily include progressive development - it only takes advantage of such variations as arise and are beneficial to each creature under its complex relations of life…If it were no advantage, these forms would be left, by natural selection, unimproved or but little improved, and might remain for indefinite ages in their present lowly condition.'[Darwin]

Which is exactly why bacteria still exist. If evolution was about hallowed ascension towards evolutionary perfection, bacteria wouldn't be bacteria anymore. That doesn't mean that the many low forms have not in the least advanced since the first dawn of life though, it just means they haven't become Greater One-Horned Rhinoceros', or Giant Clams, or *Homo sapiens*, or anything other than the bacteria they are today. They're no less important though, and their method of advancement is no less the same. It's consistent and clear:

> 'Selection acts exclusively by the preservation and ac-cumulation of variations, which are beneficial under the organic and inorganic conditions to which each creature is exposed.'[Darwin]

As are the biological foundations of the process:

'Variability is governed by many unknown laws.'[Darwin]

Well, they are now anyway. For Darwin the mechanics of variability and inheritance weren't clear, even if the results were. He was convinced that acquired characteristics were passed on through the inherited effects of the use and disuse of parts and the direct action of external conditions and found it impossible to accept that spontaneous variability could generate all the innumerable structures which are so well adapted to the habits of life of each species.

Darwin didn't know about Deoxyribonucleic acid (DNA) though, and we do. He may have been reluctant to consider the variations as spontaneous, but we don't have much choice anymore, because spontaneous variation is now transparently obvious. The discovery of DNA has opened a vast window into the mechanics of variability and shown us how characteristics are formed and how they are passed on. It has given us chromosomes and genes and shown us the very language of life itself. And our understanding increases every day. We now know that the gene responsible for controlling the depth and breadth of beaks in Galápagos Ground Finches (BMP4) also controls jaw thickness in Lake Victoria's (Africa) cichlid fish for example, and that the gene responsible for controlling speech development in humans (FOXP2), also controls the development of singing in birds. Most importantly of all though, DNA has shown us the science of genetic mutations and thus the true origin of species.

I have to admit, most mutations lead nowhere, or to unfavoured races that are eliminated by the prevailing environmental conditions. But some don't; some are neutral[*], or even beneficial, and because they don't threaten viability,

---

* Neutral variations may spread by a process called 'genetic drift'. This involves changing gene frequency that's not based on active selection.

they're capable of spreading into subsequent generations; where they meet new variations of old variations and so it goes on. Errors spread and combine and favoured races transform and evolve. Driven by competition and change the forms of life drift from one form into another until before you know it, primitive slime has become Giant Armadillos, and Lions, and Daddy-Long Legs, and Knobby Argonauts[*], and all the other species that have ever existed.

They have all evolved by a process that's so simple and elegant you don't even have to work very hard to understand it. In fact, you have to work much harder not to understand it, and people do, a lot. They work really hard to reject the evidence that's literally everywhere, and that's exactly why I'm not going to dwell on the evidence anymore. I'm not going to explain why you're 50% fruit fly, or 88% mouse; I'm not going to discuss the evolution of antibiotic resistance in bacteria, or anthelmintic resistance in parasitic nematodes; nor am I going to discuss the progressive fossil record, or the documented speciation of fruit flies. I'm not even going to discuss Man vs. Plasmodium (the malaria parasite) or the similarity of similar forms because if you don't believe in evolution now you probably never will, regardless of evidence. If you can look at the pelvis of any mammal that cannot walk, or the eyes of any creature that cannot see, or the wings of any bird that cannot fly and decide they're not evolutionary remnants, then there is, quite clearly, no point fighting. If you believe that rudimentary organs, like mammary glands in male mammals, or the vestigial legs of primitive snakes (boas and pythons), or wisdom teeth in humans, have been created for the sake of symmetry, or in order to complete the scheme of nature, then we should agree to disagree. You can cling to your particular creation myth while the rest of us marvel at reality, because change is essential and the origin of species is quite clear.

---

[*] Knobby Argonauts aren't unfortunate members of the Argonauts from Greek mythology. They're pelagic octopus instead.

What about the *destiny* of species though?
Is that also quite clear?
Yes. It is.
But only if you're objective.

Chapter 3

SUBJECTIVE

Most humans aren't objective. In fact, most don't even know what it means, so I will start this chapter by making sure that you do:

> 'objective – (adjective) 1. not influenced by personal feelings or opinions. 2. not dependent on the mind for existence; actual.' [1]

Please note, objective people aren't influenced by personal feelings or opinions. That doesn't mean they don't have personal feelings or opinions. It just means they don't believe that their personal feelings and opinions define reality. It means they can be impartial and the most fundamental part of being objective is the recognition of *intrinsic* values:

> 'intrinsic - adjective, belonging to the basic nature of someone or something; essential.' [2]

We aren't talking about what something means to you anymore, we're talking about what something means to itself and that's a very different thing. For instance, people happily kill rats without a second thought, but what does a rat mean to itself? It's impossible to say for sure of course, but the likelihood that it wants to be slowly poisoned by someone who really dislikes animals that take advantage of humans, must be zero. I suspect they're as keen to survive, and as reluctant to die, as we are and that's what I mean by intrinsic values.

It's not quite as simple as that though, because we can't assume rats, or indeed all organic beings, value themselves in the same way as we do. In fact, the likelihood that they can cope with abstract concepts like mortality is, in most cases, absolutely zero and we mustn't overcook our conclusions as a result. We can be quite sure that the purpose of existence is to survive though. And we can also be quite sure that the same applies to all organic beings, whether they're aware of it or not. And we can definitely be quite sure that life is life, regardless of the subjective feelings we may have about the vehicle that carries it.

For those who don't know what 'subjective' means, I will define that too, because it's the exact opposite of objective and a fundamental part of human emotion:

> 'subjective – (adjective) 1. based on or influenced by personal feelings, tastes, or opinions. 2. dependent on the mind for existence.' [3]

Being subjective isn't about evidence, it's about feelings, and the most fundamental part of being subjective is the *projection* of *extrinsic* values:

> 'extrinsic – (adjective) not essential or inherent.' [4]

We're not talking about what something means to itself anymore, we're talking about what something means to somebody else (i.e. X means Y to me so I will assume X means Y to X as well) and that's a very different thing. For instance, amputating the legs of old dogs has nothing to do with old dogs and everything to do with the way some people feel about old dogs (and the way some people feel about the money they can make from old dogs of course). And rat persecution has nothing to do with rats and everything to do with the way some people feel about rats. And vegetarianism has nothing to do with dead animals and everything to do with the way some people feel about dead animals. And conservation has nothing to do with extinct animals and everything to do with the way some people feel about extinct animals. The truth is extrinsic values aren't about reality, they're about feelings, and it should be perfectly obvious that feelings can corrupt perspective and warp reality. They're not objective and their subjective effects are most obvious in the domestic species. Take the Bulldog for example:

> '…who will believe that animals closely resembling the bulldog ever existed in a state of nature?'[Darwin]

Maybe you haven't really thought about it that much before, but if you haven't, think about it now: do you believe that animals closely resembling the Bulldog ever existed in a state of nature?

A Bulldog[5].

The answer must be no. I have no doubt there will be some loonies who think packs of wheezing bulldogs used to roam wild and free (hunting unicorns in the fields surrounding Atlantis no doubt), but to anyone with half a brain it should be perfectly obvious that these animals couldn't survive in the wild. They have been hideously mangled by subjective emotion and they're riddled with restrictive injurious variations as a result. For instance, the characteristic collapsed face (the sourmug) may be a defining feature of the breed, but it's also a defining threat to its health. It restricts breathing and thermo-regulation (via panting) and turns minor increases in oxygen requirements and body temperature into critical emergencies that can lead to collapse and death. In fact, they're not even supposed to go outside because of these problems. In the words of the Bulldog enthusiasts themselves, in particular those supervising bulldogsworld.com (the 'internets largest source for Bulldog information'), Bulldogs should 'avoid excess heat and be an indoor dog all the time'[6]. Which is clear evidence that continuous selection for extreme brachycephaly[*] has made trivial actions that most dogs take for granted (like moving) abnormally difficult and potentially lethal.

It's one of the reasons behind the progressive elimination of sex in the breed as well. As ridiculous as it may seem, Bulldog breeders have created an animal that can't breathe fast enough to cope with sex. Apparently, 'it is possible to free breed a Bulldog and some do it, but the stress and heat level will rise and could prove fatal'[7] and that should leave you in no doubt about what a good job these people are doing, and what a good motivator subjective emotion can be.

That's not all though, because subjective selection has also deformed its pelvic anatomy as well. It has made natural mating physically challenging, even if it could breathe, and the combined effect is a rigid barrel[†] with short legs and an inconvenient tendency to spontaneously suffocate. As a result, sex, one of the most fundamental aspects of mammalian biology, is being replaced by vaginal or trans-cervical catheterisation[‡] and even, incredibly, surgical insemination using full general

---

[*] Bulldogs are one of several brachycephalic dog breeds that are characterised by short, broad heads. Other examples include the Lhasa apso and Shih tzu and.

[†] 'Gait/Movement - Peculiarly heavy and constrained'.- Kennel Club.

[‡] A catheter is used to manually deposit sperm directly into the uterus.

anaesthesia and direct access to the uterus.

And that's *still* not all, because the same extreme head and pelvis deformations that cause problems during mating, cause problems during birth as well. They lead to monstrous puppy craniums and deformed birth canals and mean that, in the words of the enthusiasts themselves again (as a 'bulldog fact' on the bulldogsworld.com index page), 'Bulldogs require a c-section for delivery of puppies 90% of the time'[8]. *90% of the time.* Apparently, 'c-section of a Bulldog is the most preferred way and the safest!' and 'Do NOT attempt a free whelp…Be safe, not sorry! Danger…Danger…Danger!'[9]

Basically, because of a subjective attachment to *extreme* paedomorphia[*], objective selection has been hijacked and Bulldog breeders have created a dog that can't survive without emergency excitement restrictions, life-support air conditioning and routine reproductive surgery. Now, if that was their intention, fair enough, but if it wasn't, then Bulldog breeding is obviously controlled by idiots. It still involves the preservation of favourable individual differences and variations, but it's been subjectified (not a real word) and the priorities have been confused, at best. In fact, they have been beaten to death and flushed down the toilet, because it means continuous selection for morphological extremes has occurred because health is more injurious than beneficial. It means that healthy Bulldogs have been at a serious and quite remarkable *disadvantage*.

Perhaps I should repeat that:

---

[*] Paedomorphism (or juvenification as it's also known) is the retention of juvenile characteristics in adult dogs. Here's Julia Bohanna of the UK Wolf Conservation Trust (writing in the *BBC Wildlife Magazine*) to explain: 'the dog we know has evolved paedomorphic features: qualities such as floppy ears and large eyes that stimulate our need to protect and perpetuate the relationship', i.e. dog lovers often want to look after dogs that look like babies. How disturbingly pathetic is that.

Continuous selection for paedomorphic extremes has occurred because health is a serious *disadvantage* in the struggle for life.

The paradoxical effects of subjective emotional management should already be plain to see then. In the world of Bulldog breeding, ancestral races that could breathe and reproduce freely have been eliminated by the prevailing conditions (breeder prejudice) in favour of an extreme *lusus naturae* (freak of nature) that's just getting worse, and they're not alone: Pugs with granulomatous meningoencephalitis (pug dog encephalitis – an inflammatory disease of the central nervous system), Shar Peis with skin infections and Shar Pei Fever (an autoimmune disease that can lead to death), Cavalier King Charles Spaniels with brain-skull disproportions (syringomyelia) and valvular heart disease. They're all here because domestic breeds are defined by projected extrinsic emotions about superficial appearance rather than objective intrinsic assessments about the internal qualities of a dog. Favoured and injurious have been blended together in a cauldron of subjective prejudice and the result is clear: the 'survival of the fittest'[*] has become the survival of the cutest (Bulldog), or the ugliest (Bulldog), or the smallest (Chihuahua, Teacup Yorkshire Terrier), or the largest (Great Dane), or the baldest (Chinese Crested Dog), or the hairiest (Komondor, Puli), or any other cosmetic quality that generates a sentimental reaction in the adjudicating species (us).

Basically, humans 'do not and will not admire a medium standard, but like extremes'[Darwin] and the subjective consequences are clearly demonstrated by comparing the worst domestic descendents with the last wild ancestor: the Grey Wolf:

---

[*] The phrase: 'Survival of the fittest' was introduced by Herbert Spencer (in his book *Principles of Biology* of 1864), and not Charles Darwin (although he did use it in later editions of *The Origin of Species*).

A Grey Wolf[10] *and a mangled Bulldog*[11].

Grotesque isn't it. Bulldog breeders have obviously hijacked evolution and twisted the Bulldog so much it's hard to believe the two are related, and the same is true for most dog breeds. It's definitely true for the Pekingese:

A Grey Wolf[12] *and a hideous Pekingese*[13].

I don't know about you, but I would find it easier to believe the Pekingese evolved from a cushion than a wolf, and Darwin himself struggled to understand canine ancestry:

> 'I do not believe, as we shall presently see, that the whole amount of difference between the several breeds of the dog has been produced under domestication; I believe that a small part of the difference is due to their being descended from distinct species.'

Remarkably however, the whole amount of difference *has* been produced under domestication, and all dog breeds *are* the same species[*]. They have all been subjectively selected by people who think their extrinsic vision is more important than a dog's intrinsic qualities and most have suffered as a result. In fact, they have *all* suffered as a result. Out of 210 recognised dog breeds[†], not one has escaped. In the words of Mr. Chris Lawrence MBE, QVRM TD, BVSc, MRCVS and Veterinary Director of the Dogs Trust:

> 'I would not wish to give the impression that some are worse than others or that some are unaffected. I cannot think of a single breed that does not have some inherited defect.' [14]

A statement which is complimented by Mr. Bradley Viner DProf, BVetMed, MSc(VetGP), MRCVS, external advisor to Nottingham University Veterinary School and advanced general practice certificate assessor for the RCVS:

> 'We know all that breed defect stuff is not new…try naming a breed worth mentioning that does not come with its own list of potential problems.' [15]

---

[*] We will return to the issues surrounding the definition of a species later. For now we will assume a species is defined by the ability to produce fertile offspring.
[†] As recognised by the UK Kennel Club.

For example, Labrador Retrievers are predisposed to hip dysplasia (abnormal development), elbow dysplasia, juvenile cataracts, retinal degeneration and gastric torsion (stomach twists and starts to decompose). Yorkshire Terriers are predisposed to tracheal collapse, bronchitis and the partial death of its own thigh bone (Legge-Calve-Perthes disease[*]). German Shepherds are predisposed to megaoesophagus (dilation of the oesophagus), progressive posterior paresis (leads to hind limb paralysis), epilepsy, peri-anal fistulas (I will leave that to your imagination) and coagulation disorders (like Von Willebrand disease). Beagles are predisposed to hypothyroidism, dwarfism, umbilical hernias (guts end up between the body wall and the skin) and immune-mediated arthritis. Boxers are predisposed to tumours, aortic stenosis (narrow aorta) and allergies. And so it goes on. According to the Dogs Trust, and based on a Royal Veterinary College review commissioned by them, the 50 most popular dog breeds account for a staggering 322 different inherited diseases. *322.* If they were shared equally, and they're not, that would be nearly 6.5 inherited diseases per breed. Now, if that was the intention, fair enough. But if the intention was to breed happy healthy dogs, artificially forcing increasingly deformed dogs to mate with increasingly inbred relatives to 'fix' increasingly subjective morphological characteristics was just plain stupid.

FASHION

The UK Kennel Club will disagree of course, but they can say whatever they like, because dog breeding is about making different dogs, not 'making a difference for dogs'. It's about canine fashion and Crufts is the premiere fashion-based, dog mangling showcase. Apparently it helps people 'enjoy the wonderful world of dogs'[16], but Crufts isn't about 'the wonderful world of dogs'; it's about the wonderful world of dogs

---

[*] Legge-Calve-Perthes disease involves decay of the head of the femur.

that look a certain way because of artificially applied cosmetic breed standards based on the arbitrary extrinsic opinions of people who believe their own hype. It's pure fashion and anybody with any respect for nature must struggle to integrate the 'wonderful world of dogs' with the functional world of reality. For instance, 'the ridge on back formed by hair growing in opposite direction to the remainder of coat'[17] in Rhodesian Ridgebacks may be 'wonderful', but it's not functional. Given that most dogs cope pretty well without the ridge, there's absolutely no reason that the ridge should exist at all and even less reason for it to be 'clearly defined, tapering and symmetrical', especially if it might be related to the unusually high occurrence of dermoid sinus* in the breed.

The Rhodesian Ridgeback Club of Great Britain vehemently disagrees of course. In response to allegations that the ridge serves no purpose they issued the following statement:

> 'This is absolute nonsense as the ridge defines the breed from any other large brown dog.' [18]

But 'defines the breed from any other large brown dog' is an obviously superficial way of justifying an arbitrary feature, and all such justifications are the same. It doesn't matter whether it's 'high-set [tail], curled tightly over hip'[19] in the Pug, or 'hair on body moderately long, perfectly straight (not wavy)'[20] in the Yorkshire Terrier, or 'bite preferably level or slightly undershot'[21] in the Japanese Chin. They're all about the division of subjective characteristics and the only way of justifying any of them is the definition of one breed from another. Which has nothing to do with man's best friend and everything to do with man's favourite living sculpture.

---

* Dermoid sinus is a developmental abnormality involving incomplete differentiation of the spinal anatomy from the dorsal skin. It can involve a direct communication between the exterior and the spinal canal.

A Pug[22]. *Paedomorphic.*

It's utterly subjective and often has serious consequences. In the words of Mr. Bradley Viner, DProf, BVetMed, MSc (VetGP), MRCVS, external advisor to Nottingham University Veterinary School, assessor for the RCVS CertAVP (VetGP), pedigree dog enthusiast, veterinary surgeon and Royal College of Veterinary Surgeons Council election candidate:

'I'm a dog…'[23]

He decided to write an article as his own *pedigree* Bernese Mountain Dog by the way:

> 'I'm a dog and even I know that, by definition, pedigree breeding means limiting the gene pool, and with that comes an increased risk of fixing in undesirable traits', but 'if you want to keep pedigree dogs [like he does] – and it seems that a large majority do – you will just have to accept [like he has] that some degree of genetic risk goes along with the inbreeding involved.'

i.e. a highly decorated veterinary surgeon, and Royal College of Veterinary Surgeons Council election candidate, is defending pedigree dog breeding because he has (or is) a pedigree dog. And his son agrees. In an article in the *Veterinary Times* entitled *whispering my pedigree passions*[24], Oli Viner BVetMed MRCVS also defends pedigree dogs, particularly the French Bulldog. Apparently, they make him go 'just a little bit gooey' every time he meets one even though, as he himself is quick to point out, they have 'deformed faces. And bodies. And soft palates. And tails. And pretty much everything related to their appearance'. In fact, according to the enthusiasts themselves again, this time those running the 'online French Bulldog reference manual' (frenchbulldogz.org), the paedomorphic facial deformities that Oli loves so much 'are 'responsible for the adorable snorting and snuffling sounds'[25], but also for numerous respiratory health problems. These include heat stroke, elongated soft palate, stenotic nares, cleft palate, tracheal collapse and megaoesophagus and they're listed amongst a long list of other health issues and alongside the following statement:

> 'The vast majority of French Bulldogs can neither breed, conceive or whelp naturally.'[26]

Let me repeat that:

> 'The vast majority of French Bulldogs can neither breed, conceive or whelp naturally.'

Does that not strike you as a little bit…totally ridiculous?

They're the 'relative stand-up comic of canids' though and because of this Oli has no problem justifying his extreme paedomorphilia. In fact, he goes even further. He also manages to justify his educated selection of a mangled gargoyle by making the following statement:

> 'Pedigree breeds are meant to cause division of opinion,

                    The Destiny of Species

because if we left dogs free to procreate at will, we'd just have an homogenous mass of collies crossed with staffs…Now, where would the fun be in that?'

I have no idea Oli. Perhaps it might be in the dogs themselves, rather than their cosmetic appearance, but who am I to say. I can say that 'if we left dogs free to procreate at will', we would probably end up with a wolf (or a dingo), but regardless of what I have to say you have perfectly illustrated the sentimental projection of subjective extrinsic values.

Viner Senior provides the best summary however. He maintains that pedigree dog breeds are more than justified because 'the implication that hereditary diseases would just go away if breeders were more responsible is clearly nonsense'[27], and he's quite right, because wherever there is *any* breeding for cosmetic appearance there's breeding for unintended problems. Here's Caroline Kisco, spokesperson for the UK Kennel Club, on twenty years of efforts to marry physical corruption with physical health for example:

'The Kennel Club has been tackling these problems for 20 years…This isn't something you can turn around overnight.' [28]

Twenty years isn't overnight though is it Caroline. And breeding and genetic health clearly don't like each other very much. Perhaps Crufts demonstrates this better than I can though. In particular the disease predispositions of the five most recent winning breeds (most recent at the time of writing anyway):

- 2008 – GIANT SCHNAUZER. Predisposed to: Hypothyroidism, hip dysplasia, epilepsy, incontinence, autoimmune haemolytic anaemia, atopy, symmetrical lupoid onychodystrophy, crohn's disease, phalangeal cancer.
- 2007 – TIBETAN TERRIER. Predisposed to: Cataracts, Ceroid

Lipofuscinosis, Progressive Retinal Atrophy, Lens Luxation, Hip Dysplasia, Patella Luxation, Deafness, Hypothyroidism, Sebaceous Cysts, Allergies.
- 2006 – AUSTRALIAN SHEPHERD DOG. Predisposed to: Collie eye anomaly, cataracts, iris coloboma, hip dysplasia, epilepsy, elbow dysplasia, Pelger-Huet syndrome, hypothyroidism, nasal solar dermatitis, deafness, blindness, micropthalmia.
- 2005 – NORFOLK TERRIER. Predisposed to: Mitral valve disease, luxating patellas, jaw abnormalities, hip dysplasia.
- 2004 – WHIPPET. Predisposed to: Undescended testicles, heart failure, inherited eye disease.

'Undeniable' is the word, and these problems are the direct result of breeding for subjective characteristics that can only be justified because they define one breed from another. In the words of Mr. David Coffey, BVetMed, MRCVS:

> 'Anyone with even a rudimentary knowledge of the misery inflicted on pedigree dogs – and indeed cats – must be aware that the problems and dire consequences arise from the selection, by inept and ignorant breeders and judges, of animals with anatomically inappropriate, debilitating structure and physiology. The asinine world of dog breeding and showing panders to those individuals whose lives are so barren that they need to seek some limited social recognition by genetically mutilating their unfortunate canine and feline victims.' [29]

Which, rather conveniently, brings me onto the subject of cats, because they're not doing an awful lot better than dogs at the moment. 'The difficulty in pairing them'[Darwin] has been conquered by determined cat breeders with no sense and no remorse and they're quite happy to restrict freedom in pursuit of their particular aberrant vision. They're also quite happy to begin their selection with 'some half-monstrous form'[Darwin] as well, because how else could they treasure Munchkin cats? How else could these clowns decide that cats with tiny 'munchkin' legs are anything more than freak show novelties? Apparently they're no worse than Dachshunds, but that's as twisted as it is subjective. Essentially they're saying: 'dog breeders are allowed a stumpy-legged, achondroplastic (possessing deformed cartilage) dog breed so cat breeders are allowed a stumpy-legged, achondroplastic cat breed'.

For those who don't know, and in the words of the International Cat Association:

'the Munchkin is not a new mutation.' [30]

Which is probably as much as I really need to say on the matter. If they're perfectly happy to accept, no, announce that munchkins are abnormal mutations, then my work is already done.

They haven't stopped there though. Joining the Munchkin in the feline extreme breeding hall of fame is the appalling Sphynx breed, which is completely bald and so susceptible to sunburn and hypothermia that individuals shouldn't be allowed outside unsupervised; the Devon Rex breed, which is predisposed to a congenital muscle disease that regularly suffocates affected cats (due to severe spasm of the larynx – i.e. they're throttled by their own genetics); Persian and Himalayan cats which are predisposed to heritable cataracts; Maine Coon cats which are predisposed to hereditary hip dysplasia; Exotic Shorthairs with polycystic kidney disease and so on. All in the name of subjective, emotional, fashion-based human 'love'.

Sentimental suffering isn't just limited to the determined pursuit of morphological extremes either. It extends right into the heart of general animal care and zoo elephants are a prime example. Elephant keepers will vehemently disagree of course, but most of them are sadistic tyrants whose opinions generally warrant immediate dismissal anyway. According to the Animal Behaviour Research Group of the University of Oxford for example, 'the most common system used to handle female and young male Asian elephants in European zoos is free contact'[31], but that's a system that relies on the 'establishment of dominance over the elephants' by using 'physical punishment (e.g. with an ankus /elephant hook), as well as restraint and sometimes deprivation'.

That's just the beginning of the problems though, because elephants haven't evolved to live in zoos even if they aren't being mentally and physically beaten up. They haven't evolved to live in small concrete enclosures; they haven't evolved to live in foreign climates; they haven't evolved to eat high energy forage; they haven't evolved to be chained up; they haven't evolved to do tricks, they haven't evolved to change social groups; they haven't evolved to do any of these things, and they're making it quite clear. Like the dairy cows we will come to later, zoo elephant fertility is terrible, disease rates are high, and life expectancy is lower than in the logging camps of Asia. According to the Animal Behaviour Research Group of the University of Oxford again:

> 'Asian females in zoos produce…around ten times slower than rates in the wild or in extensive conditions [i.e. the logging camps of Asia]…[and] the median zoo female thus produces just one calf in her whole lifetime, compared with six in the wild …[and] the mean life expectancy (i.e. the mean age at death) of elephants in European zoos is just 15 years in Asians and 16 years in Africans, if all deaths are included.' [32]

One more time, for all those elephant 'lovers' who think elephants should continue to live in zoos:

> 'The mean life expectancy of elephants in European zoos is just 15 years in Asians and 16 years in Africans.'

Which brings me to the 'lifeboat' argument. Apparently, elephants support conservation by introducing the general public to the species and its survival challenges, but even if it's true (and it's very hard, if not impossible to quantify properly), does that matter to an elephant? It may make *Homo sapiens* feel better, but is a chronically frustrated captive elephant really thinking: 'my life is generally rubbish, but that's OK, because I know others are benefiting from my sacrifice'? Of course not. They're pretty intelligent animals, but if the general public needs educating, how is an elephant going to understand global conservation issues and the destiny of species? And even if it could, would it really be prepared to martyr itself?

I guess it's impossible to know for sure, but that hasn't stopped the utilitarian elephant 'stewards'. They maintain that the needs of the many outweigh the needs of the few and try to justify elephant captivity that way. Whatever the arguments though, conservationists have used utilitarian ethics and their 'love' of the species to justify torturing some 'flagship' individuals which is barbaric to say the least. In fact, it's intensely malicious. Effectively they're saying: 'we love elephants so much we're going to physically and mentally torture some' and the result is clear, because the zoo elephant population can't sustain itself. Their injurious conditions of life lead to high stress* which leads to high mortality and low fertility and they're clearly saying: 'if you want animals that can cope with captivity, don't keep us'.

Unfortunately, and as usual, nobody's listening. Even though high mortality and low fertility clearly indicates high stress and poor adaptation, the humans responsible continue to think that they know best and the 'compassionate' torture of zoo elephants continues. They're not alone however, and I will end this chapter with an extreme example of 'compassionate' cruelty that clearly demonstrates the effects of subjective management and delusional self-reverence.

---

* Stress refers to inadequate provisions in all areas of husbandry. It includes environmental stress, dietary stress, social stress and psychological stress (fear).

GANGOTRI

'In an act that will shock Britain's Hindu community, a veterinary surgeon escorted by three police officers [and the RSPCA]…secretly killed a cow [Gangotri] at the largest [Hare Krishna] temple in Britain.' [33]

It's true. I did.

Having been informed about the prolonged recumbancy of a cow at a religious temple I called in the Royal Society for the Prevention of Cruelty to Animals (RSPCA) and they arranged for me to openly 'secretly' kill her under police escort:

'The cow, named Gangotri, a 13 year-old Belgian Blue and Jersey cross, and much loved by the community, was killed at 9.00 am at the Bhaktivedanta Manor in Hertfordshire…Cows are sacred to Hindus, and the killing of a cow is considered to be an outrageous act.' [34]

How thoughtless of me. Perhaps there was a reason though:

'Gangotri was unable to walk.' [35]

Fair enough. Is severe bovine disability something that means very much to Hare Krishna devotees though?

'Despite being unable to walk for many months [fifteen to be precise]…the killing of a cow at a temple amounts to religious sacrilege of the worst kind.' [36]

Clearly not.

To be honest, I'm not even sure which offended me most: fifteen months of unnecessary suffering or the crazy bizarre assumption that cows are religious. Either way, I enforced my 'mistaken notion that killing is superior to suffering'[37] and slaughtered, murdered, destroyed, assassinated, euthanased, killed, 'put to sleep' (depending on your point of view[*]) Gangotri.

Needless to say the British Hindu community was shocked[†]. In fact, the world Hindu community was shocked. In their eyes Gangotri 'was sick [diseased] but had no disease [sickness]'[38] and even though I have no idea how that statement can logically exist, Mr Barry Gardiner MP[‡], and the Reverend Richard Leslie[§], and Canon Guy Wilkinson[**], and Mr. Peter Ainsworth MP[††], and Mr. Shailesh Vara MP[‡‡] all seemed to agree.

The Hare Krishnas themselves were quite sure of course. According to the head farm manager, Stuart Coyle:

> 'At no point has any vet [apart from me] stated that the pain was intolerable, merely, in the words of the official statement from the government department DEFRA, the pain was "unnecessary".' [39]

Oh well, if the pain was just 'unnecessary' that's fine.
He continued:

> 'Gangotri was unable to walk, but due to her condition there was some tolerable discomfort.' [40]

---

[*] For the record, Gangotri received a heavy dose of sedative and then a lethal dose of intravenous barbiturates.

[†] To see some of their shock please visit
http://www.youtube.com/watch?v=APY7C2j_d14 (I'm in this video) or
http://www.youtube.com/watch?v=GLm4X7j_tCs&feature=related.

[‡] MP for Brent North and the MP responsible for an Early Day Motion related to the incident (EDM 576 - Putting to death Gangotri at Bhaktivedanta Manor).

[§] Convenor of the Hertsmere Forum of Faiths.

[**] The Archbishop of Canterbury's Secretary for inter-faith relations.

[††] Shadow Secretary of State for Environment, Food and Rural Affairs.

[‡‡] Shadow Deputy Leader of the House of Commons.

Unfortunately, 'because all living things are part of God'[41], and because God, who is also Krishna, is also Govinda and 'Lord of the cows', and because 'the cow is accepted as our mother [female biological parent] in Indian culture'[42], and because 'we must always make every effort to prolong life'[43], 'tolerable discomfort' actually means discomfort which does not result in death.

Incidentally, Gauri Das, the man responsible for suggesting that we must 'always make every effort to prolong life' and one of the loudest voices in the Gangotri debate[*], has 'stepped down from his position as Temple President of Bhaktivedanta Manor after a decision by the International Society for Krishna Consciousness Central Office of Child Protection'[44]. Apparently, he was 'responsible for 'inappropriate and excessive corporal punishment' whilst a teacher at an ISKCON affiliated school in India during the 1990's'[45].

Anyway, apart from the fact that the pious leader of the justice for Gangotri campaign used to beat children inappropriately and excessively, and aside from the fact that all non-fatal 'discomfort' is perfectly acceptable to Hare Krishnas, here are the objective facts of the case:

- Gangotri was mounted and flattened by a bull in the summer of 2006.
- It's possible she had broken her back (a proper clinical examination was not possible when I saw her and no confirmed diagnosis had been made previously).
- She receives excellent nursing care and is attended regularly by Mr. Chris Day, MA VetMB, MRCVS, a homeopathic vet from Oxfordshire, as well as some conventional vets from Park Veterinary Surgery, Watford.
- Clinical inspection revealed several dressed and covered sores on the lateral aspect of her left pelvic limb and a

---

* You can see him moaning on this YouTube video:
http://www.youtube.com/watch?v=APY7C2j_d14

large area of the brisket.
- Her right pelvic limb was visibly swollen, and several dressings, including one covering the lateral aspect of the point of the hock, were also noted.
- Bandaging was evident over the carpal joints.
- Teeth grinding and abnormal vocalisation were noted on many separate occasions.
- Repeated stretching of hind limbs and rolling of the body were also noted.

Most importantly of all though, her respiratory rate was 72 breaths per minute (normal = 10-30 breaths per minute). That's between 2.4 and 7.2 times normal for a resting cow. In comparison, the normal respiratory rate for a resting human is between ten and twenty breaths per minute. It will vary between individuals of course, but whatever your personal settings, if you treble your resting breathing rate you will have some idea of how Gangotri was breathing. All that remains then is to work out why an animal with possible broken bones, definite bed sores, no exercise demands, undisputed pain and sickness *without* disease is clinically hyperventilating. Especially when you remember that overt pain responses (like teeth grinding and abnormal vocalisation) are suppressed in prey animals like cattle (because they would be advertising themselves as an easy target).

Of course, if you want to perform a proper comparison you should lie on your front for 15 months and then decide why she was hyperventilating, but in the absence of such dedication I have no doubt that Gangotri was in a lot of 'discomfort'. I also have no doubt that objective management would have ended her suffering months ago. It's also worth pointing out that her suffering wasn't just pointless, it was also against the law. Under the Animal Welfare Act 2006, all animal owners have a legal obligation to ensure protection from disease, injury, pain and suffering, even if they believe that sickness is not disease, pain is not discomfort and all suffering is acceptable as long as it's not fatal. More specifically, the Codes of Recommendations for the Welfare of Cattle state that:

> '...when an animal becomes recumbent [like Gangotri], it is important to identify the likely cause. Where there is a history of trauma, for example, falling or slipping [or being flattened by an amorous bull], a veterinary surgeon should assess the extent of any injury. Where the prognosis for recovery is poor [like a broken leg or a broken back], early intervention, by humanely destroying the animal on-farm, should not be delayed.' [46]

'However here we are [there we were] one year and quarter on [15 months later] and she was still going strong [not dead]'[47].

All in the name of subjective 'compassion'.

Anyway, I euthanased Gangotri at 9.00 am on the 13th December 2007. I decided the physical suffering of one cow was more important than the mental suffering of an entire religion, and I don't regret the anguish I caused.

Fortunately, everything is back to normal now anyway. Despite claims that 'no compensation will be adequate to address the loss of Gangotri'[48] it appears that some compensation has been adequate to address the loss of Gangotri, because the RSPCA have offered an apology and a gift of a pregnant cow and decided they shouldn't have delivered effective relief of animal suffering after all. They have rewarded the Hare Krishnas for their protests, reinforced their subjective beliefs and 'a matter that shocked them into disbelief has now been concluded on a positive note'[49]. And even if it hadn't, Gangotri had attained 'Goloka' (the spiritual world) anyway. Apparently her ashes were scattered in the Ganges and that's 'an act to bring peace to her soul'. They even flew 'the final pot of her ashes by helicopter to the Himalayan source of the river'[50] and it all really begs the question: what was all the fuss about? If she has attained 'Goloka' anyway, then her death has not affected her Karma, and the destiny of her spiritual something is something about something. Either way, she's no longer suffering and the whole episode was a perfect example of the pain and discomfort generated by people who are desperately lost in their own subjective fantasies. In this case death was genuinely not prompt, suffering was genuinely unnecessary and the minds of men had genuinely disappeared into the clouds of *ignis fatuus* (illusion). Even an obsessive pacifist like Mohandas Karamchand (Mahatma) Gandhi managed to accept the role of compassionate mercy killing in extreme cases:

> 'There is an instinctive horror of killing living beings under any circumstances whatever. For instance, an alternative has been suggested in the shape of confining rabid dogs in a certain place and allowing them to die a slow death[*]. Now my idea of compassion makes this thing impossible for me. I cannot for a moment bear to see a dog, or for that matter any other living being, helplessly suffering the torture of a slow death.'[51]

I must admit, I have no idea what he would have thought about Gangotri, but the message remains the same, because Gandhi did *not* believe 'we must always make every effort to prolong life', even if some members of society do.

And I'm not just talking about Hare Krishnas…

---

[*] Regardless of religious proclivities, this is utterly ridiculous. Killing a rabid dog not only ends its suffering, it also eliminates the risk it presents.

Chapter 4

Homo Sappiness

Gangotri was an extreme example, but the emotional impact of death isn't just a source of religious cows and compassionate torture. In fact, because it totally ignores the brutal reality of Life on Earth it's probably the source of the utopian sentimentality that defines modern natural history. Either way though, the growing cultural tendency to assume that good 'welfare' is about maximising life rather than minimising suffering obviously means that injurious variations in the domestic species aren't ruthlessly culled anymore, and that's a significant part of the paedomorphic breed mangling we discussed earlier.

It's slightly more complicated than that though, because the fashion-manglers couldn't have done it all on their own. Indeed, if their support had been limited to quack medicine and superstition most injurious variations wouldn't have lasted five minutes, regardless of emotion.

Unfortunately however, their support is *not* limited to quack medicine and superstition.

VETERINARY MEDICINE

My decision to study veterinary medicine was based on the naïve assumption that it would involve the pragmatic application of objective selection principles. I was wrong however, because veterinary medicine is *not* about the pragmatic application of objective selection principles. In retrospect, I have no idea why I thought it was, but I did and I have been pondering veterinary ethics ever since. It took me a while of course, but eventually I realised that vets are part of the problem, not the solution. Their effect is a totally logical consequence of their actions, but a totally *illogical* consequence of their intentions. In reality, saving lives may seem compassionate and righteous, but in most cases it just ends up removing the pressure to be ruthless and that just allows small problems to become larger problems that are much harder to solve.

It would be unfair to claim that the profession doesn't *end* lives as well of course, because they do. Unfortunately, this is often long after they have made a lot of money from the problem and long after the threshold necessary for objective selection. It would also be unfair to claim that the profession does not do a lot of good animal welfare work as well, because they do. Unfortunately, a lot of this work just involves supporting pointless problems (created by deranged breeders) that make them a lot of money, which is the ethical paradox I failed to resolve. For instance, *extreme* beef production has generated a cattle breed that's losing the ability to give birth. Like foeto-maternal disproportion in the Bulldog, Belgian Blue calves regularly exceed the capacity of the pelvic canal and Belgian Blue cows regularly require caesarian operations. Essentially, the breed is so extreme they can't give birth naturally and that's an injurious variation that simply could *not* exist without veterinary surgeons. By offering a compassionate alternative (i.e. the caesarean operation) the pressure to be ruthless (i.e. eliminate the problem) has been bypassed and the selection for grotesque muscular beef monstrosities has been encouraged.

Similarly, in dogs, the congenital heart defect patent ductus arteriosus could *not* have become a heritable risk *without* veterinary surgeons. In the hands of objective selection or quack medicine the deranged post-natal blood flow (between the aorta and the pulmonary artery) would have yielded a significant disadvantage in the struggle for life and the defect would have been ruthlessly culled. Unfortunately, it's not in the hands of objective selection or quack medicine; it's in the hands of the veterinary profession instead and rather than being ruthlessly culled it has been compassionately and logically encouraged.

Shar Pei entropion (inverted eyelids) only survives because it can be fixed. *Extreme* wrinkle breeding in the Shar Pei, among other breeds, often means the eyelashes rub against the cornea to cause great irritation and pain (like having lots of hairs in your eye all the time). If it couldn't be fixed it wouldn't survive, but it can, so it has[*].

A Shar Pei skin monster[52].

Dachshund spinal problems are the same. The ridiculous sausage shape may be comical and endearing, but it has stretched the spinal column beyond its physical limits and regularly leads to slipped discs and spinal paralysis (amongst other things). If these couldn't be fixed, dog elongaters wouldn't be able to elongate dogs; but they can, so they do.

I could go on and on.

In fact, I will go on and on.

---

[*] I realise I am being a bit simplistic. I have no doubt that some people would keep animals with entropion, and other injurious variations, even if they *couldn't* be treated.

Fly-strike (maggot infestation) in long haired rabbit breeds, like the English Angora and the Jersey Wooly, shouldn't be a problem at all. Rabbits with hair long enough to collect flies and maggots just don't need to exist. They're cosmetic toys and they're only here because extreme loonies find extreme fluffiness extremely satisfying. And because vets have found ways to preserve the problem. If they couldn't, cosmetic fluffiness would be eliminated by the selection pressures created by the flies, but they can so it hasn't (which actually means flies would be better at controlling fly-strike than vets).

Likewise, Lionhead Rabbit epilepsy (honestly, these breeders are trying to make a rabbit look like a lion) would rapidly disappear without treatment. If genetically inbred seizures meant they were unfavoured, epileptic Lionhead Rabbits would be eliminated, but they're not, so they haven't. They're favoured instead and the problem continues to exist.

Perhaps the most obvious example is the use of hormones to force-breed dairy cows, because that has obviously helped force-farmers to bypass normal physiological production limits and thus turned the modern dairy cow into the dying mess we will discuss at the end of this chapter. It has also made bovine pregnancy (necessary to support milk production[*]) increasingly drug dependent even though the problem is actually remarkably simple:

> 'If nourishment flows to one part or organ in excess, it rarely flows, at least in excess, to another part; thus it is difficult to get a cow to give much milk and to fatten readily.'[Darwin]

Which basically means they can't do everything. And complex veterinary solutions won't help. In fact, yet again, they actually make the problem worse. Why? Because that's how selection

---

[*] The aim is one calf per year, although this is drifting further and further as the industry uses vets to select for increasing infertility.

works. If infertility means cattle produce no offspring, then infertile variations that are struggling to cope with their conditions of life will be eliminated by definition. But, if hormonal force-breeding means infertile cattle produce *some* offspring, then infertile variations will *not* be eliminated by definition. In fact, because infertile cows often produce the most milk (which is often why they're infertile in the first place) there will be positive selection for infertility and that means force-breeding a few infertile cows has helped a small problem become a big problem that's much harder to solve. Basically, if you want animals that *can* get pregnant and milk like a waterfall, should you ever, and I do mean ever, force-breed animals that can't? No. Of course not. As Charles Darwin remarked: 'Hardly anyone is so careless as to breed from his worst animals', and yet that's exactly what's happening. Instead of recognising a blindingly obvious physiological response, veterinary surgeons plough in and force them to breed anyway. They use synthetic hormones to coerce oestrus[*] and impose pregnancy and that's the exact opposite of basic breeding logic:

> 'No breeder doubts how strong is the tendency to inheritance…That like produces like is his fundamental belief…The inheritance of good and bad qualities is so obvious.'[Darwin]

Basically, if you don't want it, don't breed from it, but still it goes on. Infertile cows are being forced to ignore their own physiology and that has, and could only have, led to a population of increasingly infertile animals. The biggest mystery is why everybody is so surprised. Milk production has rocketed and dairy cows, like every species on Earth, *can't* do everything. They're struggling with human demands and vets are facing an escalating battle with a problem they're

---

[*] Oestrus is the sexually receptive period associated with ovulation.

helping to preserve and exacerbate.

I could still go on and on.

The point is that animal mangling and the preservation of injurious variations isn't just about the people who value mangled animals, it's also about the people who support mangled animals. It's also about veterinary surgeons.

They will vociferously disagree of course, but luckily I don't need to rely on my word alone. Here's Mr. Carl Padgett, BVMS, CertCHP, MRCVS, for example, an active partner in a fourteen-vet mixed practice, current Chairman of the British Veterinary Association Animal Welfare Foundation Trustees and Senior Vice President of the British Cattle Veterinary Association:

> 'In relation to animal breeding, the veterinary profession has become somewhat blasé on these standards by accepting welfare debilitating conditions as possibly the norm.' [53]

And here's Emma Milne BVSc MRCVS, who made her name in the BBC TV series 'Vets in Practice':

> 'Vets all too commonly now will say, or write, in their clinical notes, 'This is normal' for a certain breed.' [54]

And Mr. Martin Atkinson, BVSc, MRCVS, and veterinary practice owner from Middlesex:

> 'Breeders have created the monstrosities we call pedigrees by selectively in-breeding for crippling mutations that, in the process of natural selection, would have died out as undesirable characteristics…and, all too often, with the complicity of our profession.' [55]

And Mr. Chris Lawrence, MBE, QVRM TD, BVSc, MRCVS

and Veterinary Director of the Dogs Trust:

> 'Above all else, we need to increase the awareness of
> veterinary practitioners to the consequences of confor-
> mation and disease. I don't think it's acceptable for us
> to ignore the fact that every peke and pug has noisy
> breathing because it has upper respiratory obstruction.
> And, I think – and I include myself in this – we have
> become immune to the consequences of these confor-
> mations because they are 'normal' for the breed.' [56]

I have to admit, it's hard to imagine what veterinary surgeons
are being taught at college if they're *not* aware of 'the consequences of conformation and disease', but that's what he said.

Karen Humm, MA, VetMB, CertVA, MRCVS, from the
Royal Veterinary College, went even further. According to the
Veterinary Record she said:

> 'The profession needed to take a look at itself – after
> all, it had taken a television program ['Pedigree Dogs
> Exposed' on the BBC] to bring action on the issue [of
> pedigree dog diseases], not the veterinary profession
> taking a stand.' [57]

OK, so none of them go as far as admitting joint responsibility,
but you should hear what Mr. David Coffey, BVetMed,
MRCVS, had to say about what he describes as 'the crust of
complacency that conceals our profession's approach to
animal welfare'[58]:

> 'Human society treats other animals as commodities; a
> natural resource, to be exploited for the perceived
> benefit of our kind. It is my belief that makes our
> profession a tool of animal oppression. Examples of
> our duplicity are not hard to find. In the same issue of
> *Veterinary Times* [27/7/2009], Nicky Paull [president

of the British Veterinary Association (BVA)] attempts to justify the BVA's decision to support Crufts. If ever there was a blatant example of animal abuse it is the genetic manipulation of dogs by breeders. Nicky Paull states that dog shows 'provide an excellent opportunity to educate the public about pedigree dog breeding'. May I respectfully observe that the BVA's decision suggests that it is not members of the public that need educating.' [59]

And Mr. David Bee, MA, VetMB, MRCVS, is under no illusions about the real priority:

'It seems to me that a considerable body of our profession is now driven, above all else, by the desire to obtain the maximum possible revenue from every presenting condition. This masquerades under the charade of 'gold standards'. The clinical justification is often referred to as 'defensive medicine', and the outcome of this is that clients are faced with ever increasing veterinary fees, over which they have little control.' [60]

With or without the professions acceptance however, the simple fact remains: most of these problems couldn't survive without vets, and thus one of the major reasons that we need vets, is because we have vets.

WELFARE

To be fair to the vets, most of them don't really have much choice. Most are constantly torn between the interests of the business and the interests of the animals and that's an ethical conflict that's hard to cope with (as I can testify). Indeed, I wonder if it can help explain why the suicide rate is about four times that of the general population and about twice that of other health care professionals. David Coffey, BVetMed, MRCVS, certainly thinks so:

> 'If we are to reduce the number of disenchanted members who seek solace in drugs , alcohol or death, the inconsistencies, incongruities and contradictions inherent in our profession's philosophy should be forcibly explained to prospective students.' [61]

I guess we'll never know for sure, but the suicide rate *is* high and the veterinary profession *is* confused. How do any of them justify painful procedures that have nothing to do with animal welfare for example, like embryo transfer in cattle? That's a procedure that has absolutely nothing to do with cattle health and absolutely everything to do with cattle breeders. It involves multiple hormone injections, multiple vaginal and rectal invasions and multiple cows and it has no health benefits whatsoever. It's totally about money and painful enough to warrant a legal requirement for epidural anaesthesia, which is a painful procedure in itself[*].

Perhaps we should have a look at the veterinary oath at this point, the binding agreement sworn by all members of the Royal College of Veterinary Surgeons (RCVS):

> 'In as much as the privilege of membership of the Royal College of Veterinary Surgeons is about to be conferred upon me. I PROMISE AND SOLEMNLY DECLARE that I will abide in all due loyalty to the Royal College of Veterinary Surgeons and will do all in my power to maintain and promote its interests. I PROMISE above all that I will pursue the work of my profession with uprightness of conduct, and that my constant endeavour will be to ensure the welfare of the animals committed to my care.' [62]

---

[*] The Bovine Embryo (Collection, Production and Transfer) Regulations 1995.

Which basically means: I will look after my own (and not write books that publically criticise the whole profession of course) and make animal welfare my main priority. I haven't worked out what happens when the two are in conflict yet, but given that I'm publicly announcing that the two are in conflict, I'm sure I will find out soon enough.

Incidentally, the 2008 RCVS Guide to Professional Conduct (that includes the oath amongst other things) is available online at: www.rcvs.org.uk. If you do check it, have a look out for the following phrases which clearly demonstrate the secretive and paranoid nature of the profession: 'veterinary surgeons must not: speak or write disparagingly about another veterinary surgeon'[63], 'your clients are entitled to expect that you will: uphold the good reputation of the veterinary profession'[64], 'Publicity must not be of a character likely to bring the profession into disrepute'[65], and 'veterinary surgeons must not realise they are making a financially lucrative problem worse, and, if they do, they're definitely not allowed to write a book about it'.

That last one wasn't real by the way, even if it is true.

Anyway, the point is that animal welfare is supposed to be the main priority, but is embryo transfer in cattle *really* about animal welfare?

No. It's not, and there's absolutely no way that anybody can say that it is. It's a completely unnecessary procedure in a *specifically* healthy animal. All arguments based on the manipulated production of better cows are about the welfare of farmers, and all arguments based on the need to offer the same services as everybody else are about the welfare of veterinary surgeons. In reality, this procedure may be economically justifiable, but, and this is the critical point, it specifically contradicts the oath.

And what about force-breeding dairy cows? Given that infertility is often a consequence of extreme physiological stress, it seems quite clear to me that force-breeding dairy cows specifically involves working *against* the animal's own physiology, but how on Earth is that 'ensuring the welfare of animals committed to your care', never mind the logical prevention of further infertility? It seems particularly counter-intuitive when you consider the following conclusion from a recent report by the European Food Safety Authority (EFSA):

> 'Many farmers [acting through, or on the advice of their veterinary surgeons remember] intensively manage the reproductive biology of the dairy cows by using hormonal treatments…which they perceive as economically optimal. This results in poor welfare as it deprives the animals of a coping mechanism at their disposal, to delay the onset of the reproductive process postpartum, to cope with metabolic stress caused by high production…There is no evidence that deficits in nutrition, housing, handling and management leading to poor fertility in dairy cattle can be compensated by hormonal treatments…Hormonal treatments should not be used to compensate for deficits in management.' [66]

Which suggests that the veterinary profession's involvement in force-breeding dairy cows is either greedy, or stupid.

As is it's obscene involvement in the surgical insemination of Bulldogs, because that clearly has nothing to do with animal welfare, especially when you factor in the high probability of a caesarean at the other end. Indeed, it seems pretty clear to me that getting a Bulldog pregnant by any means is cruel, but slicing into its abdomen to achieve it is just malicious. It's completely unnecessary abdominal surgery on an animal that's selfishly condemned to more abdominal surgery (Bulldogs 'require a c-section for delivery of puppies 90% of the time'[67] remember).

And what about this disturbing and hideous statement, taken directly from the RCVS Guide to Professional Conduct:

'Registration of a dog with the Kennel Club permits a veterinary surgeon who carries out surgery to alter the natural conformation of a dog, to report this to the Kennel Club.' [68]

For all those who missed it, that's 'surgery to alter the natural conformation of a dog'. Now, forgive me for asking, but how on Earth are people who 'must not cause any patient to suffer by carrying out any unnecessary mutilation'[69] also allowed to perform 'surgery to alter the natural conformation of a dog'? How? Having made animal welfare the first priority how is subjective plastic surgery acceptable?

Elsewhere, 38% don't even use post-operative pain relief following partial foot amputations in cattle. Well, 38% of a self-selecting population of UK cattle vets interested enough to complete a voluntary questionnaire on the subject anyway. I dread to think what the rest do. In fact, 3.6% of 'vets interested enough to complete a voluntary questionnaire on the subject' don't even use local anaesthetic. They just tie the cow's foot to a post, slide a serrated wire in between the two claws and saw away, without using any pain relief of any kind. And that's not all: 1.6% don't use local anaesthetic during bovine caesarean operations and 20% don't use local anaesthetic during surgery to repair umbilical hernias in calves. Here are the thoughts of Dr J N Huxley, BVetMed, PhD, DCHP DipECBHM, MRCVS, and Dr. Becky Whay, BSc, PhD, the authors of the research paper presenting these results:

'It is surprising that only 61 per cent, 68 per cent and 60 per cent of the respondents, used NSAIDs [non-steroidal anti-inflammatory drugs – a class of pain killing drugs that also includes ibuprofen] to control pain after the major surgical procedures of claw amputation, caesarean section and surgery to repair an umbilical hernia, respectively, and even then they were not administered to all the cases seen. This was in spite of the fact that they were considered the most painful procedures conducted on adult cattle and calves, with median pain scores of 10, 9 and 8 [out of 10], respectively.' [70]

'Surprising' is a pretty gentle way of describing it. I would call it disgusting.

They continue:

'Possibly of most concern is the fact that two (0.3 per cent) of the respondents stated that they did not use any analgesic agents when performing claw amputations or caesarean sections...and a further small number stated that they did not use a local anaesthetic in all cases.'

This is a group of vets interested enough in animal welfare to complete a voluntary questionnaire on the subject remember:

'One of the barriers to the provision of appropriate analgesia [pain relief] in cattle may therefore be, in some cases an unwillingness or inability to appreciate the level of pain that cattle are suffering.'

Honestly, what are these people being taught at college if they're unwilling or unable to 'appreciate the level of pain that cattle are suffering'? And how does that fit with 'my constant endeavour will be to ensure the welfare of the animals committed to my care'? *Not* using a cheap, readily available local anaesthetic product during a claw amputation, or a hernia operation, or a caesarean operation, seems to me like a deliberate way of not ensuring 'the welfare of the animals committed to their care', not the other way round.

Do you want to hear what Professor Philip Lowe, the author of a report on veterinary services for the UK government, thought about undergraduate welfare training?

'The roles, responsibilities and even the basic training
and competence of veterinarians in relation to the welfare of farm animals are unclear.' [71]

OK, so that's a specific farm animal observation, but do you
want to know how many vets in general listed 'welfare' as their
primary area of expertise in the most recent survey of the
profession (2006)?

'[blank]' [72]

That's right: there wasn't even enough to register a figure. Out
of 2747 respondents, less than fifty (1.8%), and possibly none
at all, thought animal welfare was their primary area of
expertise, even though it's the only thing they swore an oath
to defend.

Do you want to know how many listed welfare as their
secondary area of expertise?

'4%'

At least it's of secondary importance to a few of them I guess.

For those who are wondering about regulation at this
point, the veterinary profession is, rather unsurprisingly, self-
regulated. For those who are wondering what self-regulation
actually involves, here's the RCVS to explain:

'In order to practice veterinary medicine in the UK all
vets must…possess a [veterinary] degree from a university recognised by the RCVS.'

And for all those who are wondering whether that's it: yes,
that's it. Beyond a degree recognised by the RCVS, and as long
as they pay the extortionate RCVS membership fee (currently
£294), there are no mandatory Continuing Professional
Development (CPD) requirements, no mandatory clinical

competence checks and no mandatory practice standards. Indeed, according to a survey conducted by the publisher of the Veterinary Times, 40% of veterinary practices 'rarely or never' even conduct employee appraisals[73]. 40%. Basically, the only people regulating veterinary surgeons are their own clients, and they're being lucratively encouraged to believe in life at all costs. As perfectly illustrated by the arrival of kidney transplants as a legitimate treatment option. Now, I realise the RCVS thinks that 'the transplantation of kidneys in cats should be regarded as ethically acceptable in the United Kingdom'[74], but does anybody really believe that anybody can ensure the welfare of the animals committed to their care while surgically removing a kidney from a specifically healthy animal, even if there's a sick recipient, and even if the 'the level of distress caused to the source animal should be kept to a minimum'[75]? In fact, what does that even mean? The whole sentence is defined by the word 'minimum', but that doesn't actually mean anything objective. It's relative and beyond definition. In reality 'the level of distress caused to the source animal should be kept to a minimum' just means: you can hurt healthy animals a little bit, but try not to hurt them too much.

Actually, it doesn't even mean that, it actually means: you can hurt healthy animals as much as necessary. But how does 'you can hurt healthy animals as much as necessary' fit into 'my constant endeavour will be to ensure the welfare of the animals committed to my care'[76]? How are these two compatible?

They aren't, and the transplantation of kidneys is clearly just a lucrative way of using healthy cats to make large amounts of cash out of dying cats. It's a clever way of turning healthy animals into a commercial resource while simultaneously encouraging profitable sentimentality through the emotional ethical doctrine of life at most, if not all, costs.

To be fair to the RCVS, 'source animals should be used on only one occasion'[77], but given that it would be utterly impossible for an animal to donate more than one kidney and live, it's hard to imagine why they felt obliged to write that at all.

Anyway, apart from guidelines that state the blindingly obvious, the harvesting of kidneys is a perfect example of how the veterinary profession encourages the projection of subjective extrinsic prejudice. In this case there are two cats and both have the same intrinsic values. They are objectively similar and yet, based on the emotional attachments of the devoted human owners, the dying cat is valuable enough to warrant stealing part of the healthy cat.

There isn't even a legal obligation for the clients to accept and cherish the donor cat. According to the RCVS: 'source animals should only be euthanased when there is no reasonable alternative'[78], but that's just another statement that doesn't mean very much; it's more emotive waffle. As soon as you say the word 'reasonable' it all becomes relative and that means euthanasing the donor is perfectly legal.

To be honest, if you're going to do the surgery at all, it's probably the best option. There's no suffering involved in killing an anaesthetised uni-kidneyed cat. It's unconscious and then it's dead, which, if welfare is the priority, seems far better than removing a kidney then waking it up to experience totally unwarranted post-surgical pain and suffering. That's the emotional effect of death though. And fighting death is a major reason that vets exist, as Frank Busch, PhD, MRCVS, will now explain:

> 'To love an animal fully for its own sake is to treat it as an end in itself...and to press resolutely for the best interests of the animal.'[79]

Which basically means dead cats can suffer because they've lost out on life. To be fair to Frank, this particular quote has nothing to do with kidney transplants. It's actually from an ethics article about why vets and owners should be happy to consider double amputations. Citing a goat with useless front legs, a dog with just back legs, and a greyhound with no right legs it seeks to convince vets that bipedal quadrupeds are ethical and doesn't refer to potential income once.

Whatever the subject matter though, it's still nonsense. Dead animals are dead. They can't suffer because they're not alive; they can't suffer because they're dead, that's the whole point. Death is a guaranteed way of ending suffering, not causing it, and nature has been using it for 3.5 *billion* years. It's definitive and euthanasing an unconscious donor cat is clearly the most sensible option.

Actually, euthanasing the dying recipient cat and not doing the operation at all is the most sensible option, but either way, something needs to die. It's true that very few vets actually perform 'ethical' kidney transplants, but this isn't about who's performing them, this is about why they're being done at all, and why the profession's elite regards them as normal and acceptable. The RCVS is supposed to 'safeguard the health and welfare of animals committed to veterinary care'[80] but the evidence is clear: harvesting kidneys is a barbaric procedure that's profitable, canine plastic surgery is a pointless procedure that's ridiculous, embryo transfer is an economic procedure that's painful, surgical insemination is a harmful procedure that's lucrative and the sentimental preservation of injurious variations (and thus the subjective inversion of Charles Darwin's theory of natural selection) is the staple diet of the whole profession.

'COMPASSION'

Compassionate death resistance has had a pretty severe impact on the animals we 'love' then, but significantly it's not even a consistent cultural value. In fact, it's a totally inconsistent cultural value because alongside the vast death-resistance industry is a death-promotion industry that's supported by, more often than not, *exactly* the same people.

I'm talking about the meat industry of course, which is

rather unavoidably based on the early deaths of specifically healthy animals. It also provides the most graphic example of the difference between extrinsic and intrinsic values. What exactly is the difference between the *intrinsic* values of a pet dog and a meat pig for instance?

The most common response (in fact the only one I have ever received) is that meat pigs are meant to be food, but, and this is where you must really start to think about intrinsic values, what do the pigs think? Are they really thinking: 'I'm happy to die because I'm meant to be food '?

Of course they're not. That's an applied extrinsic value. It has nothing to do with pigs, and everything to do with how humans feel about pigs. In the real world, pigs value their own lives as much as dogs. As long as they're healthy, they all have 100% survival motivation and they all want to live. So what is the empirical justification for the difference between a pet dog and a meat pig?

Basically, there aren't any. They're both intelligent and neither wants to die, yet one is family and one is food. They may not be aware that they don't want to die of course, but that's beside the point. And besides, that's a similarity, not a difference and the cultural discrimination remains. In fact, many people keep pigs as pets as well, but what's the difference between a pet pig and a meat pig? Essentially, pigs are killed and cherished by the same culture, sometimes even by the same people, and that can only involve the projection of subjective *extrinsic* values.

Similarly, what is wrong with the Chinese eating dog meat? Before you answer, please bear in mind that I'm not talking about their conditions of life before death; I'm talking about the objective reasons for not eating them after death. I have no doubt that many Asian meat dogs endure appalling conditions but so do most western meat pigs, and that's not the cross-cultural issue anyway. It's the mere act of eating dogs at all that people find so offensive, but what exactly is the problem? Like the dead cat with one kidney, dead dogs can't suffer because they're not alive; they can't suffer because they're dead. They're beyond suffering and share exactly the same intrinsic value as a dead pig or a dead pet.

The point is that most people willingly discriminate between different animals based entirely upon their external appearance. Pet animals are actively kept alive based on welfare ethics that are conveniently ignored by anybody who likes to eat meat and that's not a consistent way of valuing life. There are a few people who don't eat meat of course, but I will come to them in a moment. For now it must be stated that anybody who supports the death of some animals but not others, regardless of what they look like or how they make them feel, is being hypocritical, at best. Many will stand up and proudly declare 'so what?' and that's fine. Many more will convince themselves that they're not and that's also fine. I don't care either way. I don't care whether people are honest and proud, or dishonest and delusional, but I will say this: if you can't value life based on intrinsic values, then you can't claim the moral authority and the emotional immunity needed to debate at the table of objective natural history, and that's ultimately where we're heading. You can bleat and whine at the table of subjective conservation till your heart's content, but when it comes to the politics of whaling, or the biology of sustainability, or the genetic value of hybrid tigers, those who resist and encourage death at the same time will *not* be able to claim objective consistency.

VEGETARIANS

Eliminating those that reject death while eating meat from the approaching debate, just leaves the human herbivores and I can eliminate some of them quite easily. In fact, I can eliminate about half of them with utter contempt, because about half of all people who only eat vegetables also eat fish or poultry (normally chicken).

Here are the figures:

- In 2001, the Food Standards Agency (FSA) found that 52% of 'people who only eat vegetables' eat fish, and 8% eat poultry.[81]
- In 2007, the Department for Environment, Food and Rural Affairs (DEFRA) found that 49.43% of surveyed vegetarians eat either fish or chicken or both.[82]
- And in 2009, the FSA found that 62.5% of vegetarians are 'partly vegetarian', i.e. only vegetarian some of the time, i.e. *not* vegetarian.[83]

How 'partly vegetarian' was considered as a genuine category at all beats me. Surely anybody who eats any plant material at all is 'partly vegetarian', and that must include nearly everybody that has ever lived. It *was* a genuine category though and it did include significant numbers of slightly confused botanists.

It's possible they believe fish and poultry escape the welfare concerns associated with other meat species I guess, but that's just ridiculous. Here's the Food and Agriculture Organisation (FAO) of the United Nations (UN) on fish welfare for example:

> 'Gut your fish while they are still alive if possible – the flesh will be whiter if the heart continues to pump out the blood.' [84]

That's the UN remember, and if that's what an organisation interested in implementing 'good animal welfare practices' thinks, imagine what some grizzled old fisherman can justify.

They continue:

> 'The fish need time to bleed so that the final product will have white flesh…ideally the fish should be kept cool and left to bleed for 30 mins.' [85]

Which means live gutting is done because fish consumers, which includes all pescetarians (those who don't eat land animals and birds) and about half of all vegetarians prefer their fish pumped clean of blood by its own beating heart.

Incidentally, if you aren't sure what 'gutting' means, the FAO has been quite helpful about that too:

> 'Gutting…means slitting the belly from throat to vent, removing the liver and cutting out the guts to leave the belly cavity empty.' [86]

It sounds hideous doesn't it, but it doesn't even kill them particularly fast. Admittedly, it's faster than relying on asphyxiation alone, which, according to one of the few studies available[87], takes between 55 and 250 minutes to generate death, but it still takes between 25 and 65 minutes, and that's still not particularly fast. Gutting *and* asphyxiation (the equivalent of being hung and drawn but *not* quartered) was a little bit faster, but 7-20 minutes is hardly instantaneous.

Luckily most fish are killed before they even reach the surface anyway. This is typically achieved by crushing them under the weight of thousands and thousands of their closest friends, or by dragging them along the bottom for a few hours. Here are some notes from a fishing trial conducted by the 'the UK's only cross-industry seafood body' (the Sea Fish Industry Authority) to illustrate the crushing and dragging injuries associated with being a vegetarian food source[88]. This quote concerns the state of different fish species when they're released from the net by the way:

> 'Scabbard - Dead, most of outer black skin scrubbed off…Gut inversion into the mouth was quite common. Grenadier – Dead, most of rough scales scrubbed off …Gut inversion into the mouth was quite common. Sharks – Mostly live…Belly opening and gut removal appeared to be straight forward.'

Sharks made it to the surface alive then, although that's probably a very bad thing considering the fate that awaited them there.

Perhaps carnivorous vegetarians think that fish can't feel pain. I'm not even going to honour that medieval debate with a scientific response though, because that's even worse than thinking fish are vegetables. And besides, no amount of evidence will ever convince anybody who doesn't want to accept reality. If people really believe that 3.5 *billion* years of evolution has only produced *one* species capable of feeling its guts being ripped out, or the gut-vomiting weight of countless anonymous colleagues, then they're clearly lost in a satisfying fantasy about their own self-importance, and can neither hear nor understand reason[*].

One thing's for sure, they definitely haven't been influenced by PETA's (the People for the Ethical Treatment of Animals) nauseating attempt to rebrand fish as, and it really is quite painful to write this, 'sea kittens':

> 'People don't seem to like fish. They're slithery and slimy, and they have eyes on either side of their pointy little heads—which is weird, to say the least…Of course, if you look at it another way, what all this really means is that fish need to fire their PR guy—stat…You've done enough damage, buddy. We've got it from here. And we're going to start by retiring the old name for good. When your name can also be used as a verb that means driving a hook through your head, it's time for a serious image makeover. And who could possibly want to put a hook through a sea kitten?'[89]

Obviously not people who are anthropomorphic loonies that's

---

[*] For those who want to know more about fish sentience and pain, this is as a good place to start:
http://fishcount.org.uk/fish-welfare-in-commercial-fishing/fish-sentience.

for sure, and to strengthen my case here's a genuine PETA sea kitten 'bedtime story':

> 'Sally is a Sea Kitten with attitude! While all the other Sea Kittens are washing themselves or chasing balls of yarn, Sally is busy swimming upstream to see where life will take her next. Unfortunately, years of watching her friends and family being hooked through the mouth and dragged into a harsh, alien world above have driven her mad with grief. Bitter and insane, she spends her days plotting revenge against the Land Kittens who live such happy lives in comfortable homes, free from the terror of being eaten.'[90]

Somebody, that I can only assume was an adult with a brain, actually wrote this. And PETA, that I can only assume isn't a bunch of nursery school children, actually thinks it's worth using. There's not much more to say really is there.

The 'fish and poultry are vegetables' gang clearly haven't been listening though, and for whatever reason that means about half of vegetarians immediately exclude themselves from any debate about anything ever, never mind objective debates about natural history.

To be fair, 'the oldest vegetarian organisation in the world' (the Vegetarian Society) does exclude the crazy meat-eating herbivores (by specifically stating that 'a vegetarian does not eat any meat, poultry, game, fish, shellfish or crustacea, or slaughter by-products'[91]), but even they can't escape the hypocritical label entirely. In fact, they can't escape it at all, because they still endorse the dairy cow, and she, of all animals, must surely suffer the most.

Bos tortured

Thanks to the propaganda you probably believe that dairy cows live in lush green fields and spend their days waiting to be carefully milked by buxom parlour maids with wide grins and soft hands. Apparently, there's nothing to worry about because they 'enjoy some of the highest welfare standards in the world'[92], but if that's true, I dread to think what's happening in the rest of the world, because Gangotri was living in paradise in comparison. The truth is our dairy cows are forced to work on despite the fact that they're often riddled with disease. They're force-farmed to produce as much milk as possible and because their suffering is specifically based on the fact that they're not killed, they therefore represent the final objective nail in the coffin of most subjective pro-life philosophies.

The dairy industry itself disagrees of course:

'We're proud of dairy.' [93]

And DairyCo in particular *vehemently* disagrees:

'DairyCo appoints new head to lead promotion and defence of dairy farming.' [94]

Well, they vehemently disagree with anybody who disagrees with *them* anyway. They're also the ones who will graphically explain why there's a need to defend dairy farming in the first place though, so they can simultaneously promote and defend dairy farming all they like.

In case you don't know, 'DairyCo exists to promote world class knowledge to British dairy farmers so they can profit from a sustainable future'[95]. They're 'funded entirely by milk producers, via a statutory levy on all milk sold off-farm'[96] and even though I couldn't find anything overtly positive about the industry per se (which doesn't necessarily mean it's not there), I could find an awful lot about promoting a positive image. In fact, it's one of the four main objectives of the organisation and even warrants a dedicated service department with a dedicated mission statement:

'Image management is a service designed to promote a positive perception of dairy products and dairy farming to the general public…Image management's work is all about making people feel good about milk, putting milk back on the agenda and in a positive light.' [97]

Anybody would think they're a little bit paranoid.

'DairyCo's new issues and image strategy will see greater investment of levy funds on the defence of dairy farming.' [98]

Either way, they're clearly pretty serious about promoting the industry. It's working too. According to the UK government, dairy cows produce around 13.7 billion litres annually and, given a total population of about 61 million, that's about 225 litres of milk per person per year (enough to fill a standard 80 litre bath tub nearly three times over).

A few members of the population don't drink milk of course, but not many. Indeed, Dairy UK reckons that just 1.9% of UK consumers don't eat or drink dairy products[99], and most of those probably haven't realised just how pervasive they are. They're not just limited to the obvious things like milk, cheese, butter and cream. They can also be found in baby food, ready meals, CHOCOLATE, ice-cream, crisps, cakes, pizza and many more besides. They can even be found in soap, as if bovine udder secretions have magical cleansing properties. In fact, it's almost harder to find products that don't have milk in, than products that do, and milk therefore links *at least* 98.1% of the UK population with the information that follows. And that includes a large proportion of the self-righteous, 'ethical', fish-and-chicken-avoiding vegetarian death resisters *and* a large proportion of the pious planetary shepherds who think we're here to manage all Life on Earth.

According to DairyCo then, which helps to make 'people feel good about milk' remember, and which will provide most of the evidence I need to make people feel *bad* about milk, 50% of dairy cows die before the end of their *third* lactation. *50%.* In fact, 20% die before the end of their *first* lactation and the average life span is now just five or six years (cows can actually live for 20-25 years). Pretty bad isn't it. The trouble is the dead ones are actually the lucky ones, because they're no longer tortured by the diseases of the living, like lameness:

> 'It is estimated that 25% of the national herd is lame at any one time.' [100]

That's half a million dairy cows limping right now, according to the industry itself. Some will be worse than others of course, but according to DEFRA 60,000 of them will be crippled (3% prevalence). For those who want to know what I mean by 'crippled', there are some helpful videos online at the Bristol University cattle lameness project website[*]. Score 3 describes a typical crippled cow but even these examples don't show how bad things can get.

DEFRA also reports that 1.1 million will be lame at some point each year (55% incidence) and if we combine those two figures together (incidence and prevalence) that means 1.1 million cattle are lame for 6 months of every year.

Perhaps I should repeat that, for all those who consider themselves to be ethically robust because they don't eat meat. And indeed for all those who consider themselves ethically robust because they do eat meat (such is the nature of ethics):

55% of dairy cows are lame for 6 months of every year.

Imagine the reaction if 55% of dogs were lame for 6 months of every year. Or 55% of polar bears, or tigers, or any other wild animal you can think of. There would be outrage. Because it's cows though it's actually perceived as 'normal', particularly to

---

[*] www.cattle-lameness.org.uk/Practise-your-Mobility-Scoring.

those who are supposed to be looking after them, as the UK government is well aware:

> 'The main stumbling block in the control of lameness appears to be related to the lack of awareness of the lameness level on farm by both herdsman and the veterinary profession. There is evidence that veterinary surgeons underestimate the level of cattle lameness more than herdsman.' [101]

Let's return to the DairyCo disease figures though. This time it's mastitis:

> 'A recent study by DairyCo…suggested the average incidence rate was 71 cases per 100 cows per year.' [102]

That's 1.42 million cases per year, and if we combine these *industry* mastitis and lameness figures together that means 2 million cows suffer from *at least* 2.42 million cases of disease every year (i.e. some will have one or both more than once). I say 'at least' because mastitis and lameness isn't a comprehensive list, as I'm sure you can imagine. Indeed, according to a variety of other industry sources, 400,000 suffer from uterine infections every year (20% incidence)[103], 60,000 suffer from left displaced abomasums* every year (3% incidence), 120,000 suffer from Milk Fever† every year (6% incidence)[104], and who knows how many have chronic rumen burn and constant diarrhoea‡. Basically, if we combine just these five diseases, 2 million dairy cows will suffer *at least* 3.1 million

---

* Left displaced abomasum (LDA) is a potentially fatal condition involving deranged physiology and abnormal displacement of one part of the stomach. This physically obstructs the gastrointestinal tract leading to severe production losses and, if it isn't treated, to chronic starvation and death.
† Milk Fever involves deranged calcium metabolism and is related to high milk production.
‡ The diet modern dairy cows must eat to support milk production often causes excess acid production in the rumen and chronic diarrhoea.

cases of disease every year, and if you share the incidence risk equally, each individual has a 155% chance of getting sick every year.

Perhaps I should repeat that, for all those who think life should be preserved at all costs:

> If you share the incidence risk equally, each individual dairy cow has a 155% chance of getting each year.

Dairy cows may 'enjoy some of the highest welfare standards in the world'[105] then, but not being the lowest doesn't mean their *actually* high.

Hayley Campbell-Gibbons, the Chief Dairy Advisor of the National Farmers Union (NFU), disagrees of course:

> '...the dairy industry has made significant progress in improving welfare standards over the past 20 years.' [106]

But I wonder what the Animal Health and Welfare panel of the European Food Safety Authority (EFSA) thinks, particularly regarding what it calls 'the major welfare problem for dairy cattle':

> '...there has been no reduction in the prevalence of lameness in the last twenty years.' [107]

And I wonder what DairyCo thinks:

> 'The incidence of lameness has increased significantly since a farmer based survey in the late 50's found animal incidence to be around 5% [currently 55%].' [108]

I guess Hayley doesn't mean lameness, but if that's the case, what exactly does she mean?

Maybe she means longevity in general, but if, as Hayley suggests, 'dairy cow longevity has been steadily increasing over the past decade'[109], why does the EFSA think 'the longevity of dairy cows has declined greatly over the last 40 years'[110]. They can't both be right so who is selectively interpreting the truth? Do you think it's the trade association that's financed by farmers, or an independent scientific panel specifically tasked by the European Commission to objectively assess the welfare risks associated with modern dairy farming?

Unfortunately, I have no legally watertight idea, but if we're interested in twenty years of improving welfare, it's definitely worth having a look at DairyCo's perspective:

> 'In reality few cows exit dairy herds free from a disease problem…When reasons for culling are examined, the vast majority of cows leaving dairy herds do so for health reasons…While age and yield account for 25% of all cows leaving dairy herds in this survey, it is likely that there is an associated health problem…'[111]

And then the comments of Mr. Roger Evans, a director (and former chairman) of First Milk and treasurer of Dairy UK:

> '…many modern dairy cows just aren't lasting long enough. Fertility, lameness and mastitis take more cows out of productive life than culling for performance or management.'[112]

Both of which are supported by the conclusions of the EFSA:

> 'Culling occurs almost always because of the perceived malfunctioning of the cow involving the cow's difficulties in coping with the housing and management system or other aspects of its environment…'[113]

Which is all fairly self-explanatory really isn't it.

Do I really need to point out, again, that the DairyCo quotes are freely available on a website dedicated to 'making people feel good about milk', and that all arguments will be based around the possibility of improvements in the future despite having achieved worse than nothing in the past?

Anybody fancy one last 'reassurance' from Hayley?

> 'Importantly, consumers can be confident that over 95% of British dairy farmers are operating to Assured Dairy Farm Standards, which sets and monitors the highest standards of animal health and welfare.' [114]

I am confident they're are operating to Assured Dairy Farm Standards Hayley; that's the problem. Here's a typical health and welfare standard for example?

> 'Lameness and mastitis are health problems that have long been associated with dairy cattle and whilst assurance cannot eliminate these conditions completely, it requires farms to adopt proactive health planning.' [115]

Hardly binding is it, and it means that 100% of cows can be lame or mastitic and the farm's quality can still be assured. In fact, because none of them contain any limits to the amount of any diseases, quality assurance schemes are actually defined by minimum legal standards, not maximum welfare possibilities, and all they really 'guarantee' is that farms aren't illegal.

No wonder the industry is so 'proud'.

Cattle vets are just as proud of course. Here's John Blackwell for example, the current president of the British Cattle Veterinary Association (BCVA):

> '…if things are so bad how many prosecutions of dairy farmers have there been over the past 12 months?' [116]

How casual is that though? Essentially, John, the president of the BCVA remember, assesses current welfare based on past prosecutions, but that's as careless as it is defensive. And besides, and as he knows full well, it's difficult to prosecute farms that have the full support of the supervising vets (i.e. the people that make their living from them).

In case you aren't convinced about veterinary loyalty, here's the BCVA Junior Vice President and long term member of the Endell Veterinary Group Farm Animal Department, Keith Cutler, BSc, BVSc, MRCVS:

> 'Most of the farmers I see care very deeply about their animals' welfare.' [117]

How lovely. If only he hadn't said it after openly admitting that 'the extreme Holstein cow is an inbred breed, and fertility, mastitis and lameness are big problems'.

Loyal vets can say whatever they like though, because it's quite clear that the extrinsic desire for walking udders has yielded an animal that's in physiological meltdown. More importantly, it's also quite clear that those who drink milk must either support death (if you do want to end their suffering) or torture (if you *don't*) and that brings us back to the subject of consistency, because, as previously mentioned, objective natural history debates can only include those who *don't* resist and encourage death (or torture) at the same time.

Please note, I'm not excluding everybody who drinks milk. I'm just excluding everybody who wouldn't happily shoot the entire national dairy herd to *improve* welfare.

Please also note that I don't have a problem with dairy farming per se. I have seen how well it can be done and I would love to see it more often. The problem is good dairy farms are few and far between and nowhere near enough to justify the public's 'positive perception of dairy', or indeed the self-righteous preaching of pious lacto-vegetarians.

That just leaves the vegans then who *can* claim genuine consistency. Unfortunately however (and quite bizarrely), this consistency is often based on the behaviour of Gorillas. Apparently, Gorillas don't eat animal products so humans shouldn't, but humans aren't Gorillas are they. That's like saying antelope don't eat warthogs so lions shouldn't; or hermit crabs don't eat ducklings so foxes shouldn't; or pickled onions don't eat wildebeest so crocodiles shouldn't. And besides, Gorillas *do* eat animal products. Despite claims to the contrary, their diet includes a significant amount of snails and insects and that means they're not, and we can ignore, vegans.

In summary then, objective natural history debates can only include those who:

- *don't* think humans should behave like Gorillas.
- *don't* think fish and poultry are vegetables.
- *don't* think death is right and wrong at the same time.
- *don't* think death can *cause* suffering.

And, ultimately:

- *don't* think their gushing sappiness defines reality.

And with that said, let battle commence.

Chapter 5

RACISM

There's no way of knowing how many 'stewards' are omnivores unfortunately, or how many are strict herbivores, or indeed how many are confused pescatarians and pious lacto-vegetarians, but the probability that they all value life consistently is almost zero. Indeed, they don't all value life consistently, because almost all of them support killing healthy *wild* animals even if they don't support killing healthy meat animals. In fact, they're always killing something to preserve something else but why are some lives ever more valuable than others, and why on Earth is it OK to kill Pollack but not Cod?

For those who don't know, the worldwide Cod population in recent decades has crashed and the 'stewards' are now demanding that we eat Pollack instead. They're overtly commanding us to kill some and not others, but what exactly are the objective differences between the two, and how can death be right and wrong at the same time? Perhaps celebrity chef Hugh Fearnley-Whittingstall can explain:

'We need to find alternatives to threatened fish – buy pollack instead of cod.' [118]

That's a good point Hugh. I will consider extinction in the next chapter though. For now we must ignore the differences between the last individual and the rest, and focus on the objective differences capable of justifying life and death at the same time. Perhaps the United States National Oceanic and Atmospheric Administration National Marines Fisheries Service (I guess that must be the USNOAANMFS but I'm not 100% sure) can explain the difference:

'Atlantic Pollock…can be distinguished from cod by their greenish hue.' [119]

Is that it? Are the 'stewards' really telling us to kill Pollack instead of Cod because Pollack have a 'greenish hue'?

Yes. They are. To be fair, there is an 8 inch difference in maximum size as well, and several pounds difference in maximum weight. And Pollack are 'slightly paler on the belly' of course. And let's not forget that the chin barbel/whisker thing is 'small' in Pollack and 'distinct' in Cod. Other than that though, the most obvious difference between Pollack and Cod is, according to the United States National Oceanic and Atmospheric Administration National Marines Fisheries Service, 'their greenish hue'.

Basically, they're a slightly different colour and that's the same difference that justified human slavery for thousands of years. It's blatant racism. Inter-species racism rather than cross-cultural racism I admit, but they're both descendants of the same common ancestor and there's just no other way of describing the fatal persecution of one race over another based entirely upon the way they look.

Remember, I don't care who kills what. I'm not trying to tell anybody how to behave, or trying to claim that any one living thing is more or less valuable than any other. I'm just trying to separate subjective from objective by demonstrating how superficial appearance can make death right and wrong at the same time, even though, in the words of Professor Richard Dawkins:

'There exists no objective basis on which to elevate one species above another. Chimp and human, lizard and fungus, we have all evolved over some three billion years by a process known as natural selection.' [120]

There are plenty of *subjective* reasons for elevating one species above another though, and the wilful persecution of Grey

Squirrels is another perfect example, because that's almost entirely based on the fact that they're not red. Apparently, Grey Squirrels need to be killed to protect Red Squirrels, but again, the only real difference is their colour.

A Red Squirrel[121] *and a Grey Squirrel*[122]. *Or is it the other way round?*

The Red Squirrel Survival Trust (RSST) disagrees of course:

'Red Squirrels…have a bushy tail and ear tufts.' [123]

But is 'a bushy tail and ear tufts' really enough to justify life and death at the same time?

'We agree with the view that targeted culling is essential in order to secure the survival of red squirrels.' [124]

Yes, apparently it is, although I should point out that there is also a 'humane' caveat:

'…we prefer non-lethal methods of control.' [125]

But I'm sure the RSST don't advocate 'non-lethal methods' when it comes to the meat they consume. Either way, they're quite clear about the reasons for culling:

> 'Grey squirrels are an unwanted and invasive species in the UK.' [126]

But it's definitely worth finding out what they mean by 'unwanted and invasive'. Maybe the European Squirrel Initiative (ESI) can explain:

> 'The grey squirrel was introduced to England, Scotland and Ireland from North America in the Victorian era.' [127]

I guess that's a deliberately introduced invasion, but whatever it is, Grey Squirrels are here because of humans and that makes all subsequent consequences our fault rather than theirs. And besides, the only real difference, other than minor deviations in tail bushiness and aural hairstyle, is still just colour. It's still about appearance, rather than intrinsic value, and that's *not* objective natural history.

It's more than that though, because it also involves the persecution of a favoured race and that's the exact opposite of objective natural history. Here's the European Squirrel Initiative to explain why Grey Squirrels are so awful for example:

> 'In Britain, it has few natural predators. It has successfully adapted to British lowland conditions. It is omnivorous, breeds strongly and is an aggressive settler equally at home in urban parks and the countryside.' [128]

But those are observations that should generate respect, not 'targeted culling', and anything else specifically means prejudice *against* success in the struggle for life. Which small squirrel do you think the 'stewards' would be culling if the situation was reversed for example? If the native squirrels were grey and the deliberately introduced 'invasive' squirrels were red, which would be sacred and which would be 'unwanted'?

I should point out that Grey Squirrels are also being racist, but I doubt whether it's actively premeditated, and I have absolutely no doubt that it's *not* motivated by colour. In fact, I should also point out that racism is a fundamental part of natural selection (because selection is selective by definition). The difference is that natural selection is unconscious racism based on function, whereas anthropogenic selection is conscious racism based on appearance.

Before we leave the squirrels, I must quickly mention the Red Squirrel Protection Partnership's views on squirrel taxonomy, which should be assessed with the British Natural History Museum's total global species estimate of 10-100 million in mind:

'There are over 267 million different species of squirrel in the world.' [129]

Now, I'm sure the RSPP has just made a genuine typing error, but given that the difference between red and grey is also the difference between 'mystical' and condemned to death, I really can't resist hoping they haven't.

Anyway, in case you aren't convinced that cosmetic racism and prejudice against success define modern conservation, here's the World Wide Fund for Nature (WWF) to explain why the lives of some jellyfish are extrinsically worthless. In particular those of a comb jellyfish species that was introduced into the Black Sea by human activity:

'They eat both zooplankton, the food of commercially important fish in the Black Sea, and the eggs and larvae of the same fish species.' [130]

i.e. the WWF doesn't like comb jellyfish because they *do* like 'commercially important fish', even though these jellyfish are, like Grey Squirrels, just taking advantage of an opportunity *we* gave them:

> 'The comb jellyfish arrived on ships from the American Atlantic coast in 1982…[and] by the mid-1990s, they accounted for 90% of the total biomass in the Black Sea - a biomass more than the total annual fish catch around the world.' [131]

That's not worthless though, that's utterly amazing. It's pride-swellingly successful and for all those who respect all Life, rather than some lives, comb jellyfish should be a source of great wonder.

The US National Science Foundation (NSF) disagrees of course:

> 'Because jellyfish reproduce quickly, are hardy and face few competitors or predators in many degraded waters, they can quickly overrun and dominate ecosystems.' [132]

But apart from being a near perfect description of human success as well, that just confirms that jellyfish are amazing and that the only real reason to vilify them is *success* in the struggle for life. It's counter-evolutionary, but it's a view that's been endorsed by the conclusions of a paper that combines the environmental expertise of Australia's national science agency (the Commonwealth Scientific and Industrial Research Organisation), the Ecology Centre of the University of Queensland, the Rosenstiel School of Marine and Atmospheric Sciences at the University of Miami, the Institute of Environmental Sustainability at Swansea University and the Department of Biodiversity and Conservation Biology at the University of the Western Cape:

> 'Mounting evidence suggests the structure of pelagic

ecosystems can change rapidly from one that is dominated by fish…to a less desirable gelatinous state.' [133]

I would suggest that 'less desirable' is probably a matter of perspective. That's the conclusion however. Apparently, jellyfish swarms are awful because they're taking *advantage* of human impact, but how does that work? Surely taking advantage of human impact is a good thing?

> 'Human induced stresses of overfishing, eutrophication, climate change, translocation and habitat modfication appear to be promoting jellyfish…blooms to the detriment of other marine organisms…with lasting ecological, economic and social consequences.' [134]

I guess not. If you prefer 'other marine organisms', and if you think that humans are the centre of the Universe, I guess that jellyfish success is a bad thing.

Cane Toads face a similar prejudice, even though they have exactly the same intrinsic mission as any other animal. They're not worthless everywhere of course, because everybody is quite happy with them in their native locations, but as soon as they escape their value completely inverts and this amazingly expedient species instantly becomes a villified loathsome 'pest'. Australia in particular really resents their invasion, but is it really their fault?

> 'They were deliberately introduced from Hawaii to Australia in 1935, to control scarab beetles that were pests of sugar cane.' [135]

Clearly not.

Incidentally, 'it turned out that the Cane Toad couldn't actually eat the cane beetle' [136] anyway. Apparently 'the beetles lived in the tops of the sugar cane stalks and the toad couldn't jump high enough to reach them', which would be pretty funny even if it wasn't true.

I digress though. The real issue is the active discrimination against yet another species specifically because it's succeeding and specifically by people who claim to represent natural history. Cane Toads have exploded from a population of just 100 in 1935 (imported by the, rather fascinating Australian Bureau of Sugar Experimental Stations) to more than 200 million today and they should be a target of cymbal-crashing respect, not golf-club-wielding hatred. They're fulfilling the aims of natural selection, and they're doing it specifically *because* of the relentless impact of humanity, as the International Union for the Conservation of Nature (IUCN) well knows:

'It thrives in degraded habitats and man-made environments, and…to date no effective controls have been implemented.' [137]

Basically, Cane Toads are completely ignoring everything humans throw at them, openly mocking the 'stewards', and, like the Grey Squirrel and the comb jellyfish, they should fill all objective Naturalists with genuine respect.

Even if they are 'ugly'.

FASHION, AGAIN

It should be quite clear by now that death is right and wrong at the same time because color and shape are more important than consistency and function, and that means we're talking about fashion again. Even though life is, for all objective purposes, life, the 'stewards' value some lives more than others based on the same sort of subjective morphological love that justifies pedigree dog breeding.

Indeed, they have even started their own version of pedigree breeding. The Royal Bengal Tiger has been split into four distinct breeds for example (I simply can't bring myself to list them). I should point out that this has already started to precipitate a whole new crop of unnecessary genetic deformities (the list already includes shortened tendons of the forelegs, club feet, kidney problems, deformed spines, congenital cataracts, psychological disorders, and *bulldog* faces characterised by a snub nose, a domed head and wide-set eyes), but the racism hasn't stopped there, because there's even active discrimination between superficially identical tigers, as the Zoological Society of London (ZSL) will now explain:

> 'The vast majority of captive tigers are known hybrids or of unknown origin and so are not useful for conservation breeding purposes. Only tigers whose single subspecies ancestry can be traced back through written records can be included in conservation breeding programmes.' [138]

Why though? Why are known hybrids and unknown pedigrees openly excluded from 'genetic lifeboat' communities and specifically labeled as 'useless' for conservation purposes? Apparently each subspecies has evolved to live in a specific place, but so have some humans and nobody would describe hybrid humans as 'not useful'. Well, nobody who is sane anyway, and particularly not those who work for the Zoological Society of London I'm sure. Nobody would even describe them as 'hybrid' but that hasn't stopped active tiger racism. Essentially, conscious prejudice has breached the boundaries of morphological racism and is now running wildly down the corridor of genetic racism flapping its big crazy wings of subjectivity at the same time.

It's not just limited to tigers either. Apparently, and according to the IUCN's Red List, 'cross-breeding with domestic dogs represents a significant threat to the long-term persistence of dingoes'[139], but, apart from the amazing suggestion that successful breeding is a 'threat', does anybody really think dingoes care whether they're 'pure' or not? In fact, given that 'many hybrids are indistinguishable from pure dingoes', how could they care? In fact, if the only real way to tell a worthless 'hybrid' from a priceless 'pure-blood' is 'skull morphometrics' and 'DNA tests', why on Earth do we care, why on Earth is cross breeding a 'threat', and why on Earth is there 'a need to assess the genetic make-up of dingo populations throughout their distribution in order to identify the prevalence of hybrid forms and inform dingo conservation planning.'?

I'll tell you why: it's because these painful experts have lost themselves in the complicated world of their own subjective racism. They want 'pure' dingoes because they're 'stewards' and they like purity and they know best. End of story. This is conservation after all, and conservation is about managing nature, not observing nature (natural history).

Incidentally, dingoes are feral descendents of domestic dogs anyway, as the IUCN is well aware:

> 'Austronesian-speaking people transported the dingo from mainland Asia to Australia and other islands in southeast Asia and the Pacific between 1,000 and 5,000 years ago.'

Which makes their managemental megalomania even more ridiculous.

Anyway, most overt racism still focuses on external appearance, rather than genetic purity, and it sometimes even divides the racists themselves. Recent calls by the Songbird Survival Trust (SST) to fight uncontrolled songbird predation by killing magpies have really upset the apple cart for example. They think that 'magpie predation of nests and fledglings is having a considerable impact on the reproductive capacity of other birds and their numbers must be controlled'[140], but the RSPB disagrees. They don't believe there's any 'evidence that control of crows and magpies is necessary for the recovery of songbird population', and Chris Packham, of the BBC's Natural History department, went even further:

> 'The trust's reasoning comes down to the same old misinformed chestnut - that evil magpies are causing the decline in smaller songbirds. It's kneejerk ornithological racism.'[141]

Rather convenient use of the term 'racism' I have to admit, but he doesn't stop there:

> 'It's true that some magpies prey on the nests of smaller birds during the breeding season, but this is for perhaps three or four months of the year and only affects young birds that are easily replaced. The magpies never kill the more valuable breeding adults.'

Unfortunately, he has failed to realise that fighting magpie control with statements like 'young birds that are easily replaced' and 'more valuable breeding adults' is just using age discrimination to fight race discrimination and that's clearly just more subjective kneejerk ornithological prejudice.

Magpies are a native species by the way, that have been preying on songbirds for thousands of years, which is probably why the RSPB doesn't support a cull. Paradoxically however, they do support 'legal control when it is necessary to prevent serious damage to agriculture or as part of gamebird management'[142] and they do 'control foxes and crows at Abernethy Forest in Strathspey, where...predation resulted in low productivity of the increasingly rare capercaillie'[143]. i.e. the RSPB only supports the culling of predators in situations where there's a threat to *rare* birds (capercaillie) or *common* birds (pheasants etc.). Which is, at the very most, confusing and consistent at the same time. Perhaps David Hoccom, a spokesman for the RSPB, can explain it better:

> 'We do not think that trapping and killing magpies is justified in most situations. In certain circumstances ...it may be necessary to reduce magpie numbers.'[144]

Clearly not.

The SST's magpie cull is supported by plenty of others though, including, rather predictably, some 'concerned RSPB members'. It's even supported by the Countryside Alliance who are…actually, I don't care who they are. Basically, the song-bird vs. predator debate is a big squabble over which birds different 'stewards' like the most. It is, quite frankly, a lot of playground bickering and perfectly illustrates the inconsistent valuation of life and the subjective nature of conservation.

The importance of colour and shape has even inspired a conservationist competition. Here's Rebecca Aldworth, the Director of Humane Society International Canada, to *try* and explain:

> 'We are closer than ever to stopping cruel commercial seal hunts, including Canada's annual slaughter of harp seals. But we need your help!' 'We need every EU citizen to let decision-makers know that they support a complete and unconditional ban!' [145]

'Every EU citizen', wow. How are you going to achieve that Rebecca?

> 'The EU resident who gathers the most signatures …will win a trip to Canada's stunning harp seal nursery next spring.'

I see. You're going to bribe people, with a competition to win a holiday, just because you really like Harp Seals. I should point out that some of these Harp Seals are maliciously tortured during the hunt, but that's not the issue. This is about killing seals at all and I can find no objective reason to sympathise with dead animals, never mind an objective reason to make the following promise:

> 'Help us keep the promise I made to these animals on

my very first visit to the ice - that one day they will be able to live out their lives in the peace and beauty of the pristine Canadian ice floes.'

Yes, you read it correctly; Rebecca Aldworth promised a load of seals eternal peace. I'm not entirely sure what she plans to do about Polar Bears and Killer Whales of course, but I assume she has thought of something, because she won't be able to protect the lives of millions of 'innocent' seals otherwise? I'm also not entirely sure what crime they're innocent of, but that's beside the point isn't it, because they're 'innocent' nonetheless and they clearly deserve to be removed from the peace-threatening reach of apex predators.

What about the 'innocent' fish the 'innocent' Harp Seals will kill though Rebecca? Why don't the fish that *won't* be eaten (by the seals that could be killed) deserve to live out their lives in the peace and beauty of the pristine ocean *beneath* the Canadian ice floes?

'…Thank you for all your doing to save the seals.' [146]

I'm not doing anything to save the seals. I'm trying to find out why the fish the seals will eat don't deserve to be saved:

'…Ban the cruel seal trade.' [147]

I guess Rebecca just doesn't like 'sea kittens' all that much.

Remember, I still don't care who kills what. I'm just trying to reveal the racist foundations that support modern conservation. I'm just trying to show that the whole philosophy is based on extrinsic attachments to external appearance and I will end this chapter with a simple comparison that will severely test your ability to remain objective. It will challenge all preconceived beliefs because if colour and shape aren't enough to justify life and death at the same time, why is size?

WHALES > CHICKENS

Without reinforcing Charles Darwin's 'fanciers like extremes' stereotype, and without disagreeing with Richard Dawkins' 'species are objectively equal' statement, why, exactly, is it OK to kill chickens but not whales?

Before you answer, please bear in mind that almost all general death-resistance philosophies have already been undermined (by dairy cows and the vegetables formerly known as fish and poultry) and that just leaves us with the objective differences between a whale and a chicken. Admittedly whales are quite large, to say the least, but only if you forget that life is life, regardless of the way it looks. The International Fund for Animal Welfare (IFAW) may think that 'the soulful, timeless grace and presence of whales, represents the very mystery of nature itself'[148] for example, but why doesn't the soulful timeless grace and presence of *chickens* represent the very mystery of nature itself? I have to admit, I think that every-thing that has ever lived represents the very mystery of nature itself, but either way there's absolutely no reason to discrimi-nate between animals that are different, and even less reason to kill one while simultaneously worshipping another.

Everybody who really likes whales will vehemently disagree of course, probably using a response similar to that of pedi-gree dog breeders:

This is absolute nonsense as differences define one animal from any other animal.

But 'defines one animal from any other animal' is blatant cosmetic discrimination and blatantly can't objectively explain why individual whales are more important than individual chickens.

What can then? Perhaps we should crunch the numbers before we reach a firm conclusion:

- WHALES - According to the International Whaling Commission (IWC), and including all catches taken under objection, all aboriginal subsistence whaling catches and all special permit catches, the total number of whales killed in 2007 was: 1931.
- CHICKENS - According to the Food and Agriculture Organisation (FAO) of the United Nations (UN), the total number of chickens killed in 2007 was: 50 *billion*.

Why, exactly, was it OK to kill 25 million chickens but not one whale then? Actually, let me put it a bit more objectively:

How is 1 > 25 million?

Perhaps your resistance to whaling is about genuine welfare concerns, rather than morphological racism. Perhaps you just can't understand how it's possible to kill a moving target from a moving boat with a moving harpoon, and you would be more than justified. Here's an extract from the 2005 Annual Report of the International Whaling Commission (IWC) to demonstrate why welfare is a concern (as well as why politics is funny) for example:

> 'Australia noted that the instantaneous death rate for minkes [in the Japanese hunts] varies between 35-44% but that the instantaneous death rate for Norway's commercial whaling is 80%. Japan expressed irritation that the same questions were asked every year.' [149]

Here's another one (from the IWC's Annual Report in 2003):

> 'Japan restated their belief that this Agenda Item [Collection of Animal Welfare Data] was inappropriate for this Workshop…The Japanese delegation left the room.' [150]

There have been enormous improvements of course, but it's still not enough to eliminate welfare concerns completely. The whalers may be trying, but even the best whaling statistics from the most concerned whaling nation show that 20% survive entry of the harpoon (see the first IWC quote above) and that's more than enough to justify genuine welfare concerns.

Is it enough to exceed the suffering of 50 *billion* chickens though? Chickens may have a better quality of death (if we forget that most are suspended upside down by their ankles before being killed, and that those birds that are shackled are just the 'lucky' ones who weren't trampled to death in the shed or suffocated during transport etc.), but their quality of life isn't even in the same league. A whale spends its life in the wide open ocean doing natural whale things in a natural whale environment, whereas a commercial chicken spends its entire life in a windowless shed with 30,000 of its closest friends and a space allocation that makes a shoe box look spacious.

In fact, chicken environments are so cramped it's worth having a proper look at the figures to facilitate the comparison. I will have to concentrate on the UK broiler chicken industry for obvious reasons, but that's still over 830 million birds a year. Here's the National Farmers Union (NFU) to explain how wonderful everything is then:

> 'In the poultry meat sector, the EU Broiler Welfare Directive has recently been introduced and will be implemented by DEFRA in the UK in June 2010. This sets legal maximum stocking densities – which the Red Tractor scheme already more than meets.' [151]

The legal maximum stocking density defined by the EU Broiler Welfare Directive is 42kg/m^2 by the way, or **21 birds/m^2**, or 476cm^2/bird. For those who need a familiar reference point, a sheet of A4 paper is 624cm^2, i.e. if every 2 kg bird was allocated its own luxurious A4 sheet of paper to stand on, there would be **16 birds/m^2**.

Let's see what the NFU means by 'more than meets' then shall we:

'This [Red Tractor] Assurance Scheme does not permit planned stocking regimes which exceed 38kg/m^2.' [152]

'Assurance' indeed, and definitely 'more than meeting'. 38 kg/m2 is still **19 birds/m^2** though, despite UK government recommendations[153] of 34 kg/m^2, or **17 birds/m^2**. It's definitely 'above and beyond legal requirements' though, so we can't really complain. We still haven't breached an A4 sheet of paper yet though. Maybe the RSPCA can do it, with 'the only UK farm assurance scheme dedicated to improving the lives of farm animals'[154] ('Freedom Foods'):

'Stocking density…must never: exceed 30 kg/m2 [**15 birds/m^2**].' [155]

Brilliant, the RSPCA freedom food scheme allows each bird slightly more than a sheet of A4 paper. Not exactly 'freedom' though is it. In reality, each RSPCA monitored bird does have an extra 190.5 extra cm^2 (13.8 x 13.8 cm area), but that's still just the size of a CD case and nowhere near enough to claim 'freedom'. In fact, almost all debates about stocking densities and the welfare of chickens are pointless anyway, because they're always limited to the inclusion or exclusion of a few extra cm^2. In reality, the world's oceans represent 'freedom', not 666.7 cm^2, and a whale's quality of life vastly exceeds that of a chicken as a result. Which brings me back to the original question:

How is 1 > 25 million?

Perhaps you think the two are unrelated, and that whales aren't about chickens. You would be absolutely right of course, because whales *aren't* about chickens, but natural history isn't about

 The Destiny of Species

hypocrisy either and all dead animals are just dead. For instance, what is the intrinsic difference between this whale[156]:

and this chicken[157]?

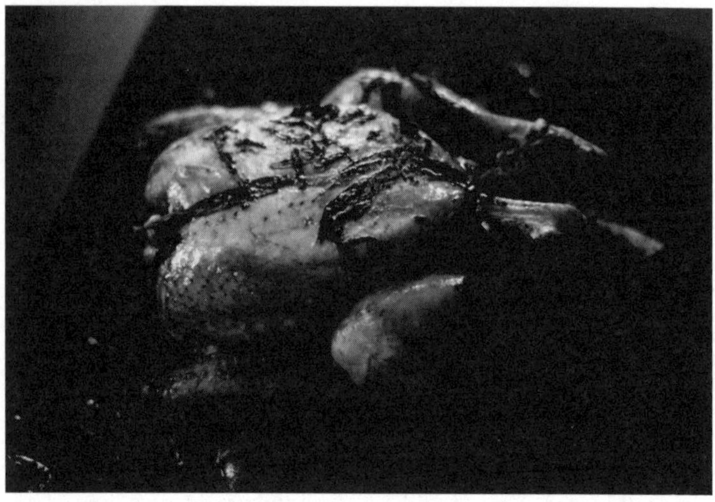

And what is the intrinsic difference between either of them and this mole[158]?

Or this Buffalo[159]?

Or this…thing?

Or this fish[160]?

Or even these carrots[161]?

What are the intrinsic differences between any of them?

There aren't any. They're all dead (or vegetables) and they all have exactly the same absence of *intrinsic* value. If we're being objective and consistent whales aren't about chickens, but dead animals aren't about inconsistent anthropomorphic sympathy either.

Perhaps you just don't like the Japanese very much. Perhaps you represent a much more orthodox form of racism. They do seem to be public enemy number one, even though there are other nations involved. Norwegian whalers have it easy in comparison for example (maybe because they actually manage to kill a lot of whales at the first attempt). And aboriginal subsistence whalers are positively encouraged. Apparently, 'subsistence whaling – such as that practiced by Alaska Natives and in Russia – is fundamentally different from the resumption of commercial whaling being pursued by the Government of Japan'[162], but I doubt whether the harpooned whales would agree.

That quote is from IFAW by the way, who are supported

in this view by, amongst others, Greenpeace and the WWF:

> 'Greenpeace does not oppose subsistence whaling by indigenous peoples.' [163]

> 'WWF recognises the human need for aboriginal subsistence whaling.' [164]

What they have all failed to realise is that subsistence whaling is fundamentally different because it's fundamentally worse, not fundamentally better. Here are the thoughts of a rival conservationist group for instance (which IFAW has subsequently joined), this time an anti-whaling coalition led by the World Society for the Protection of Animals (WSPA):

> 'Killing methods used during ASW [Aboriginal Subsistence Whaling] hunts are recognised to be less accurate and efficient than those used in commercial whaling operations, resulting in longer times to death (TTD), lower instantaneous death rates (IDRs), and higher 'struck and lost' rates...' [165]

It should be pointed out that 'WSPA does not contest the right of aboriginal peoples participating in whale hunts as sanctioned by the IWC, international and national laws'[166], but the point is everybody who objects to commercial whaling while supporting subsistence whaling is condemning and supporting the 'painful slaughter of whales'[167] at the same time. You could call it totally inconsistent, but as we have already seen numerous times before, that's nothing less than completely typical.

Anyway, after all of that, and armed with the knowledge that some conservationist groups don't object with whaling in principle, while others do except if the killing is worse than 'inhumane', how is 1 > 25 million?

Basically, the only way anybody can make one whale more

valuable than 25 million chickens, or one whale more valuable than *one* chicken, or indeed one anything more valuable than one anything else, is subjective cosmetic racism. It's all about superficial differences, rather than intrinsic similarities, and that is *not* objective natural history.

Life is life after all, regardless of how we may feel about it.

Chapter 6

EXTINCTION

If life is life, regardless of how some lives make us feel, what is the objective difference between the death of *an* individual and the death of the *last* individual, and why are the 'stewards' so worried about Lonesome George?

For those who don't know, Lonesome George is the planet's last surviving Pinta Island Galápagos Tortoise. There are plenty of other Galápagos tortoises, but George is the last one from Pinta Island and he's become a conservation icon as a result.

Why exactly is he so special though? Perhaps the Charles Darwin Foundation can explain:

'The Pinta tortoise is one of the smaller species.' [168]

I see. Surely his divinity involves slightly more than just a difference in size though:

'He has a 'saddle-backed' shell or carapace.' [169]

OK. I assume he's uniquely small and 'saddle-backed'?

'Smaller 'saddle-backed' types are found on Española or Pinzón [as well].' [170]

Oh. Clearly his morphological sanctity has nothing to do with being exceptionally small and unusually 'saddle-backed' then. Maybe Charles Darwin can explain the difference:

'Captain Porter has described those from Charles and from the nearest island to it, namely, Hood Island, as having their shells in front thick and turned up like a Spanish saddle [saddle-backed]…M. Bibron, moreover, informs me that he has seen what he considers two distinct species of tortoise from the Galapagos, but he does not know from which islands. The specimens that I brought from three islands were young ones; and probably owing to this cause, neither Mr. Gray nor myself could find in them any specific differences.' [171]

It would seem that even Charles Darwin couldn't separate Galápagos tortoises definitively at first glance, even if the Charles Darwin Foundation can.

Incidentally, here's what the Charles Darwin Foundation thinks about the Pinta Island subspecies:

'It is a poignant tale illustrating the devastation that followed the arrival of humans to the shores of Galápagos. Hunting tortoises for meat greatly affected the numbers of tortoises on Pinta Island.' [172]

And here's what Charles Darwin thought about the Galápagos Tortoise in general:

'The breastplate roasted…with the flesh attached to it, is very good; and the young tortoises make excellent soup.' [173]

A 'poignant' comparison as I'm sure you'll agree.

The Charles Darwin Foundation is also a member of the Alliance for Zero Extinction by the way, but apart from the preposterous existence of an alliance for *zero* extinction at all, we will return to Charles Darwin's views on extinction later.

Oh alright then, here's a little taster:

'On the theory of natural selection, the extinction of old

forms and the production of new and improved forms are intimately connected together.'

Anyway, the Galápagos Tortoise subspecies have been divided by slight morphological variations that weren't immediately obvious to Charles Darwin. I have yet to work out why such trivial differences are enough to separate tortoises and not humans, but that's the reality. George is sacred because he's very slightly superficially unique. And because he *might* be very slightly biochemically unique as well. Here are…some scientists to explain why there's some doubt:

> 'We explored the complex evolutionary history of the G. becki PBR [Puerto Bravo] population on Isabela Island by analyzing variation at ten nuclear micros-tellite loci relative to a genotypic database including 354 individuals from all extant populations of Galápa-gos tortoises…The nearly extinct G. abingdoni on Pinta was included in this reference population database for the first time by way of genotypic data collection from six museum specimens. Bayesian clustering revealed a widespread pattern of mixed ancestry in the PBR population. Of particular note, one of eight PBR individuals with a G. hoodensis (Española)- like haplotype (PBR03) exhibitted a strong signature of G. abingdoni ancestry and an assignment to the Pinta population (q-value = 0.743).' [174]

I hope you're keeping up. I know it's a little bit esoteric, but this is what modern species conservation is all about:

> 'PBR03 falls in the centre of the PBR–Pinta F1 q-value distribution. Combined with the results from previous mtDNA analyses, these data suggest a hybrid origin of PBR03 resulting from a mating between a G. becki female from PBR with a G. hoodensis (Española)-like

haplotype and a male from Pinta.'

Which is, as I'm sure you'll agree, marvellous news. For those who gave up, PBR03 is half Pinta Island Galapagos Tortoise and I'm sure he's very proud.

'Unfortunately PBR03 is a male.'

Oh. He won't be helping Lonesome George (♂) to save the subspecies then, not unless one of them learns how to sexually invert anyway. Indeed, George has even failed to breed successfully with female tortoises (of another subspecies), so the chances of him breeding successfully with another male are even more remote. I guess he just doesn't care about breeding anywhere near as much as the 'stewards'. This isn't about George though is it, this is about conservation, and the Charles Darwin Foundation isn't going to give up the fight that easily:

'Cloning is theoretically possible.' [175]

Yes, you read it correctly. They're even considering cloning George and that clearly has nothing to do with George and everything to do with the way George looks. Apparently, 'it would be a true shame to see the end of another unique species of Galapagos tortoise'[176], but George isn't intrinsically unique; he's just very slightly superficially unique.
And extremely rare of course.

LESS > MORE

Lonesome George is the most extreme example, but the escalating value of decreasing numbers is a fundamental part of modern conservation and it's all based on the racist values we discussed in the previous chapter. In fact, having skipped over the sustainability issue previously it might be worth returning to the Cod vs. Pollack debate once again. That's entirely based on the escalating importance of dwindling numbers, even though the difference between the two is, for all objective purposes, just a 'greenish hue'.

Richard E. Grant, Emilia Fox, Terry Gilliam, Lenny Henry, O.T. Fagbenle and Greta Scacchi will all disagree of course, but anybody who thinks cuddling a dead cod with no clothes on is a show of ethical compassion is clearly mad. For those who don't know, these celebrities have all taken part in a photographic campaign to promote sustainable fishing by posing nude with a dead Cod, but apart from the fact that fish don't understand sustainability even when they're alive, if you're going to sustainably cuddle anything it should be a live cod that's subsequently released, not a dead Cod that couldn't care less.

Either way, if the intrinsic values of live animals aren't defined by the way they look, and the intrinsic values of dead animals aren't defined by the way we feel, are Cod lives really more valuable than Pollack lives? Or does this entire campaign rest on racist concerns about saving Cod for our children, and for our children's children, and for our children's children's children etc. etc.? Yes. It does. And that is, without a shadow of a doubt, anthropocentric conservation rather than objective natural history.

That's fine of course, but I do need to point out that Cod aren't here for our children, or our children's children, or any other member of our genetic legacy. They're here for them-selves. I also need to point out that dead individuals aren't emotionally affected by the fate of the species either way. They don't care whether there are none or loads left when they die. They're just dead, and preserving the last few because we have separated individuals from species (more on this later), and because we really want our genetic descendants to be able to eat them, just devalues the billions that have already died and confirms that the 'stewards' value scarcity over abundance. It means that killing Pollack isn't about Cod, it's about us.

The 'stewards' will claim they have a right to life of course, but who exactly is 'they'? Do they mean the Cod that have been killed, or the Cod that haven't been killed, or indeed the Pollack they're proposing to kill? I guess they mean the species as a single unit in time and space, but we will come to the separation of death from extinction in a moment. Right now it should be perfectly obvious that the importance of live and dead individuals has become a relative function of what they look like and how many remain, and that is, at the very best, totally inconsistent. In case you aren't convinced, here's Nigel Dougherty BVSc, MRCVS, to explain why individual Takahēs[*] are more important than they were:

> 'When your numbers are down to the colloquial handful, it is easy to see how every living Takahē counts.' [177]

The blatantly obvious implication being:

> When your numbers are not down to the colloquial handful, it is easy to see how every living Takahē doesn't count.

And the blatantly obvious conclusion being:

> 'All animals are equal, but some animals are more equal than others.' [178]

It's all a function of external appearance and relative abundance, and conservation scientist Holger Kreft, a post-doctoral fellow in the Division of Biological Sciences at the University of California, has even found a way to measure less as greater than more:

> 'Normally you want to…protect a maximum number

---

[*] The Takahē is an endangered flightless bird from New Zealand.

of species…but you also want to focus on unique species which occur nowhere else.' [179]

Remember, unique species aren't intrinsically unique, they're just cosmetically unique:

> 'To capture that uniqueness, Kreft *et al* at UC San Diego, the University of Bonn and the University of Applied Sciences Eberswalde used a measure of biodiversity that weights rare species more than widespread ones.'

For all those who missed it, that's 'a measure of biodiversity that weights rare species more than widespread ones', i.e. less > more. Do I really need to say anything else?

For those who remain sceptical, here is former WWF ecologist Esmond Bradley Martin on why individuals are more important when they're rare:

> 'I supported the use of saiga antelope horn as a substitute for rhino horn from the early 1980's. In my opinion it was the correct policy at the time. But I stopped around 1995, when I read about the start of the sharp decline in saiga populations.' [180]

Basically, through Esmond, the WWF (which is 'for a living planet'), was recommending the slaughter of saiga antelope to preserve the rhinoceros because less is more valuable than more, but had to change this policy in 1995 after realising that saigas had became less abundant than more and thus more valuable than less.

In fact, less abundant is probably a gross understatement, because saiga numbers have fallen by, according to the WWF, a staggering 96%:

> 'Saiga antelope populations numbered over one million

as recently as the early 1990's, but have been reduced to no more than 40,000 in total…The major initial decline was caused by selective poaching for horns for use in traditional Asian medicine.' [181]

They vehemently deny their role in promoting saiga 'horns for use in traditional Asian medicine' of course:

'WWF never actively promoted saiga antelope hunts.' [182]

And they have even pinned the blame on poor old Esmond to try and wriggle clear:

'Dr. Esmond Bradley-Martin is referred to and cited as being a WWF ecologist [in the New Scientist article]. While Dr. Bradley-Martin did do extensive consultancy work for WWF, he has never been a full time employee of the organization. The views he expresses in the New Scientist article are therefore his own, and do not represent the views of WWF.' [183]

But it's definitely worth having a look at the recommendations of the 1990 IUCN/SSC African Elephant and Rhino Specialist Group (AERSG) Status Survey and Conservation Action Plan at this point:

'All nations in which rhino products are consumed should actively promote the substitution of alternative animal products such as water buffalo or saiga antelope horn.' [184]

And then it's definitely worth working out which organisation that 'never actively promoted saiga antelope hunts' actively funded this report:

'The costs of the meeting [where the report was drafted] were largely met by [wait for it…] the WWF.' [185]

I guess that's inactive activity, but whatever they think, they *were* involved in 'actively promoting saiga antelope hunts' and thus turning less value into more by turning more numbers into less.

ALL = ONE

The point is that modern conservation is based on the fact that species are much more important than individuals, and that's a subjective difference that definitely warrants closer examination. Here's another quote from the BBC Wildlife Magazine for example, this time from one of Sophie Stafford's *editor's letters* (April 09):

> 'Sadly, this iconic seabird is facing an uncertain future. Like the Polar Bear, whose tribulations are regularly depicted on tv, the puffin faces the same insidious threat of climate change, though its troubles occur beyond the cameras out at sea.' [186]

Apart from the characteristic doom and gloom however, this article has used singular terms like 'iconic seabird', 'polar bear' and 'puffin' to describe entire species. The quote suggests 'the puffin' and 'the polar bear' are facing an 'insidious threat of climate change' as single units, but species are defined by individuals, not the other way round. They're collections of mortal beings that matter to themselves and turning them into anonymous collections based on the way *they* look and the way *we* feel ignores the constant 'struggle for life'. It separates the death of individuals from the death of species and means that, like 'the saiga' and 'the rhinoceros', 'the puffin' and 'the polar bear' aren't defined by the reality of death anymore (i.e. puffins and polar bears are always dying), they're defined by the possibility of extinction instead (i.e. the *last* puffin and the *last* polar bear might die) and that's *not* objective. For instance, Birdlife International, CEE Bankwatch Network, Climate Action Network Europe, European Environmental Bureau, European Federation for Transport & Environment, European Public Health Alliance - Environment Network (now known as the Health and Environment Alliance (HEAL)), Friends of the Earth, Greenpeace, International Friends of Nature and the WWF (together known as the 'green10') are all quite right to state that:

'Extinction is forever.' [187]

But so is death, and all individuals must die at some point. And besides, when has 'the puffin' ever faced a *certain* future? The truth is that living seabirds and polar bears and indeed all living creatures have never, and will never face a certain future.

I do feel obliged to add a little disclaimer here, because I can't guarantee that I've managed to shake the 'all = one' habit entirely. In fact, it's so pervasive throughout culture and conservation that I have only just realised it happens, and only just realised how hard it is to stop. Indeed, I have done it on purpose with 'the dairy cow' at the end of Chapter 4.

I *will* continue to dissect the phenomenon though because it's crucial to understanding why managing nature (conservation) is not observing nature (natural history). I have to admit, I don't have a problem with it per se. I just don't think it should be used to demote the 'struggle for life' that every creature faces every day, in favour of the 'struggle to exist as a unique cosmetic form' that species face according to subjective extrinsic observers.

The WWF disagrees of course:

'The orang-utan, the 'man of the forest', is severely threatened.' [188]

'The mighty polar bear is under threat.' [189]

And IFAW has managed all sorts of crazy nonsense:

> 'By taking direct action on IFAW campaigns, you will
> be giving hope to thousands of animals around the
> world that suffer from…the risk of extinction.' [190]

The most relevant nonsense is the assumption that extinction
applies to species rather than individuals. To be fair, IFAW
may actually be using the term 'animals' in a truly individual
sense, but that just means 'direct action on IFAW campaigns'
somehow involves promising immortality to individual animals
around the world. And if they're not using it that way 'direct
action on IFAW campaigns' somehow involves promising
immortality to individual species, who aren't genuine units
even if they could understand the 'hope' they're being offered.

IFAW is the welfare charity that objects to whaling *except*
if it's worse than inhumane remember. Which rather conveniently brings me back to the sustainability of whaling. Don't
forget, some conservationist groups don't object to whaling in
principle; they just object to whaling that might lead to the
death of the last whale:

> 'WWF strongly opposes commercial whaling, now and
> until WWF is convinced that the governments of the
> world have brought whaling under international
> control, with a precautionary and conservation-based
> enforceable management and compliance system
> adhered to by the whaling nations.' [191]

> 'Greenpeace is campaigning to end all commercial
> whaling. Commercial whaling has never been sustain-
> able. Greenpeace does not oppose subsistence whaling
> by indigenous peoples.' [192]

OK, so Greenpeace doesn't think commercial whaling can ever
be sustainable, but what about slightly profitable subsistence

whaling, as openly practiced by the indigenous peoples of Greenland? They hunt Minke Whales, Narwhals and Finn Whales as aboriginal subsistence whalers and then sell whatever they can't sell locally to the commercial company Arctic Green Food. Who then turn the meat into a variety of delicious products, like 'minke whale mix', steak, minced meat, lightly salted blubber and flipper/tail among others, ready for commercial sale to the general public[*]. So, what about slightly commercial whaling?

I suspect Greenpeace would eventually reach a bizarre conclusion where some dead whales are acceptable while others are not, but what exactly are the intrinsic differences between sustainably harvested dead whales and unsustainably harvested dead whales, irrespective of whether they're commercial or not? What do you think this Bowhead Whale[193] thinks about being sustainably hunted for non-commercial reasons for example?

Nothing. It's dead.

Incidentally, the main target whale species was sustainably harvested in 2007 anyway. Minke Whales made up 81.5%

---

[*] For more details please visit www.arcticfood.gl.

(1574 of 1931) of the total catch and according to the IUCN Red List:

> '…there is no indication that the global population [of Minke Whales] has declined to an extent that would qualify for a threatened category.' [194]

And besides, even if they weren't common there are still no objective reasons for elevating them over any other species (never mind the humble but spectacularly marvelous chicken). Anything else is just another conservationist squabble over when, where, how and by whom something is OK to be something but only if it isn't something else:

> 'All animals are equal, but some animals are more equal than others.'

Before moving on, I must mention 'the tiger', because 'the tiger' that everybody refers to today does not include a single individual that defined 'the tiger' when intensive conservation efforts began in 1973 (Project Tiger). They're *all* different individuals even though they're all treated as *one* conservation unit.

There's no doubt 'the tiger' is under threat by the way, but 'the tigers' will never stop being under threat, irrespective of conservation efforts. They will *always* live for a maximum of 10 -12 years (ish) and never stop being individuals here for themselves rather than a species here for us. That hasn't stopped the conservation army working themselves into a frenzy though. According to the WWF there are less than 4000 tigers left in the wild (although there are an appalling number in captivity, particularly in America) and the loss of this particular morphological entity, as a species, rather than as individuals or genes, has become a global conservation priority that consumes millions and millions of dollars.

It's not even working. After 36 years of Project Tiger there are less tigers now than when it started. In India for example, and according to the Project Tiger website[*], there were 1827 Royal Bengal Tigers in 1972, and 1411 in 2008. Which is a 23% reduction, and hardly a resounding success. The 'stewards' will claim that things would be an awful lot worse if it wasn't for them of course, but for whom? All the dead tigers who were alive 36 years ago? All the people who have been uprooted by conservation authorities? All the deer that weren't killed by a smaller tiger population? I guess they mean the 1411 tigers currently alive, but they *will* all die at some point and they're now so reliant on conservation that the species as a whole has been reduced to a shadow of its former glory. These animals give no quarter and ask for none in return and yet the 'stewards' have stepped in and turned a proud hunter into a pitiful victim, which is obscene. I don't blame the tigers of course; they're just taking advantage of whatever opportunity arises. I do blame the 'stewards' though, for degrading this, or indeed any of life's soldiers with their subjective racism and hypocritical sympathy.

Some have even turned tigers into novelty circus acts. The Wildlife Conservation Society offers people the chance to 'see tigers do puzzles'[195] for instance, but where's the respect in that? That's not objective natural history, that's ritual humiliation.

Captive tigers are unlikely ever to return to the wild either, despite the assertions of the Zoological Society of London (ZSL). They reckon 'zoo tigers support their wild relatives'[196] but apart from the fact that supporting their wild relatives is definitely not their choice, nobody has ever reliably proved that it's actually possible to release captive tigers, as the ZSL themselves are well aware:

> 'It is unlikely that any tiger born and raised in a zoo will ever be suitable for release.'

---

[*] The Project Tiger Website can be found at www.projecttiger.nic.in

They do go on to add:

> 'Should the time ever come, cubs born to zoo tigers in natural habitat enclosures and raised there, with access to live prey and minimum contact with humans would be much more suitable.'

But that just helps them hide behind future possibilities so they can post-rationalise their decision to turn ruthless hunters into wretched cash-cows. Indeed, Heather Sohl, Species and Trade Officer at WWF-UK, believes 'there is no legitimate reason for captive breeding, particularly as reintroducing tigers into the wild has never been proved possible'[197]. To be fair to Heather, she was referring to the captive breeding of tigers in China to feed China's fruitcake obsession with grinding animals into non-medicinal medicinal dust, but if she believes captive breeding elsewhere is perfectly acceptable then she's just another painful hypocrite.

Save the Tiger Fund has this to say:

> 'There is also precious little evidence that zoo or farm-bred big cats can ever be reintroduced successfully into the wild.' [198]

Which is supported by the assertions of the Global Tiger Patrol:

> 'If the wild tiger became extinct, most experts agree that it is extremely doubtful whether it could ever be reintroduced.' [199]

Clearly Save the Tiger Fund and the Global Tiger Patrol haven't spoken to the Zoological Society of London though. The Zoological Society of London also justified genetic racism against 'hybrid' tigers earlier remember.

The point is that captive tigers are captive for us, not them, and their incarceration is even more disturbing than the descent of wild tigers into the shame of 'conservation-reliance'. It's appalling and I for one will not be joining 'the tiger' degradation effort. I will celebrate them as fellow soldiers of Life instead, and, regardless of whether they're currently alive or already dead, I will respect them for what they represent, rather than who they rely on.

A Tiger, that hasn't asked to be 'conservation reliant'.

In fact, let's put things in perspective shall we, because the cultural significance of 'the tiger' has absolutely nothing to do with its evolutionary significance in reality, and that's not objective natural history. For instance, according to the British Natural History Museum:

'Some 30,000 million species are estimated to have evolved during the last 545 million years. Only 10-100 million are still living today, so the majority of species that have ever lived on Earth are already extinct.'[200]

And that means 'the tiger' represents between 0.00001% (1/10 million x 100) and 0.000001% (1/100 million x 100) of species alive today, and just 0.0000000033% (1/30 billion x 100) of species that are estimated to have existed in the last 545 million years. Which should make you question the inflated significance of 'the tiger', or the deflated significance of everything else, depending on your point of view. I maintain it's the latter and so did Charles Darwin:

> 'The most humble organism is something much higher than the dust under our feet; and no one with an un-biased mind can study any living creature, however humble, without being struck with enthusiasm at its marvellous structure and properties.'[*]

The 'stewards' are even happy to vilify poachers even though most of these people are poor displaced locals who are more concerned about staying alive than the subjective demands of condescending tourists and comfortable conservation campaigners. In fact, if the 'stewards' had to struggle for life and someone offered them umpteen times their annual income to kill an animal that they may, or may not be aware is endangered, they'd at least think about it, and rightly so, because a dead tiger has value. Demented, superstition-based loony value[†] I admit, but value nonetheless and the condemnation of poaching can only involve people who are privileged enough not to have to and stupid enough not to realise.

THE LUXURIOUS THRONE OF PRIVILEGE

To illustrate the cultural racism that has been generated by the separation of death from extinction, I will use a personal anecdote about a nice lady I met in Nepal's Royal Chitwan National Park. She was in the park as a wildlife tourist on the

---

[*] Charles Darwin. *The Descent of Man* (1871)
[†] The demand for tiger parts almost entirely involves bizarre cultural twaddle with no basis in reality.

trip of a lifetime and she was captivated. She was positively effervescent about the jungles and the wildlife within and quite certain the whole lot should be conserved forever.

Her true colours soon became clear however. She may have been positively effervescent about Nepalese wildlife, but she was negatively stagnant about the wildlife of Surrey. Apparently the local deer had been eating the flowers in her lovingly manicured garden and she wasn't in the least bit impressed. For her they were a 'damn nuisance' and I couldn't help being punched in the face by her gross duplicity. From her luxurious throne of privilege she was commanding the local Nepalese population to find ways of accommodating tigers and rhinos, but under no circumstances should deer be eating the flowers in her garden. In her twisted, racist mind boundary populations should accept a constant threat to their livelihoods and lives while 'eco' wildlife tourists should *not* accept a constant threat to their Chrysanthemums and Snapdragons.

It doesn't stop there though. This middle-class 'steward' is a mere flea in the biosphere (a drop in the ocean), because there are many more with their own luxurious thrones of privilege. Indeed, nature loving conservationists are always being mildly inconvenienced and disproportionately irritated by herons in the fish pond, or aphids in the rose garden, or squirrels in the orchard, or nettles in the flower bed, or moles in the croquet lawn. For instance, in March 2009 the BBC Wildlife Magazine defended and condemned poisoning in the same issue. Apparently it's perfectly acceptable to poison slugs if they threaten your 'sunflowers', but it's perfectly *unacceptable* to poison White-Tailed Eagles if they threaten your livestock.

The slugs aren't even killed quickly. Here's Julie Roberts, editor of BBC Gardens Illustrated and author *Slug Deterrents*, to explain:

> 'The pellets contain Ferramol, a stomach poison. Once the pellet has been ingested, the slug immediately stops feeding and dies three to six days later. Unlike

other pellets…the slugs can move away to die, leaving the treated area slug-free.' [201]

This is a magazine that aims 'to inspire readers with the sheer wonder and beauty of nature and help them understand and appreciate the truly amazing world around them'[202] by the way. In fact, in the very next issue (April 2009) there's an article entitled *Respect the Cockroach*. Clearly the 'beauty of nature' and the 'truly amazing world' includes everything but slugs.

To those with a consistent vendetta against untamed nature (like most farmers), I can only say, fair enough. But to those who expect millions of people who live on less than a dollar day to resist temptation and revenge while they whine like spoilt children about mole hills, slugs, jellyfish and a whole host of other slightly tiresome disturbances in their otherwise luxurious and privileged lives, I can only say that you make me ashamed to be human.

Chapter 7

EXTANTION

Before we go any further, it's now time to decide whether you have *really* accepted Charles Darwin's theory of natural selection. In particular you must decide whether selection (inequality) can ever involve equality (non-selection). Let me put it another way: is natural selection fair?

No, of course it's not. Selection is selective by definition and that means evolution, and thus the fundamental basis of life on Earth, is inherently *unfair*. The 'preservation of favoured races'[Darwin] simply can't occur without the elimination of unfavoured races and thus it's vital to accept inequality if you're going to represent objective natural history.

This is where things start to get really interesting then, because conservation is specifically about the judicial preservation of unfavoured races and clearly doesn't represent natural selection. In fact, because saving species that aren't favoured by the prevailing conditions can only involve prejudice *against* natural selection, conservation is actually the *exact* opposite of natural selection.

The 'stewards' will claim that humanity has changed all the rules of course, because we're responsible for changing the prevailing conditions, but the prevailing conditions are *always* changing and something is *always* responsible. It's nothing new and anthropogenic change is no different. And besides, that would involve totally missing the point anyway because it should be quite obvious by now that life is life regardless of how many similar lives remain.

The corruption of Charles Darwin's theories hasn't stopped there though, because the widespread belief that species deserve time to adapt must also involve completely misunderstanding the whole point as well. The BBC Wildlife Magazine is trying to work out how to 'protect species from environmental change'[203] for example, but is natural selection ever about *protecting* species from change?

No. Of course it's not. It's about eliminating those that don't fit and preserving those that do and it never ever involves the 'protection of species from environmental change'. In fact, the 'protection of species from environmental change' is the complete opposite of the whole point because species aren't distinct entities that must be preserved as sacred anyway, they're transient collections that are constantly fragmenting instead.

Jeffrey McNeely, the *chief scientist* of the IUCN, disagrees of course:

> '…given the success of these species in adapting to continually changing conditions, perhaps we should replace our feelings of superiority with a bit more humility and simply make sure that we are not bringing about changes so fast and extreme that they will not be able to keep on adapting.' [204]

But apart from forgetting that species only appear well defined because countless numbers of intermediate forms are extinct (I will return to this shortly), and apart from suggesting that being a 'steward' is *not* a feeling of superiority, the *chief scientist* of the IUCN has completely forgotten, or never actually known, that species *don't* evolve themselves in the right direction; they're unconsciously evolved in as many directions as possible and the *right* direction picks them.

Perhaps I should repeat that, because this is a ubiquitous misunderstanding that has nothing to do with reality or Charles Darwin's theory of natural selection:

> Species do *not* evolve themselves in the right direction; they're unconsciously evolved in as many directions as possible and the right direction picks them.

In fact, Jeffrey is so far away from what Charles Darwin actually said I can only assume that he, along with the vast majority of his fellow 'stewards' and 'chief scientists', hasn't even read what he describes as the 'most important book ever published on biology'. If he had he would know that Life hasn't survived for more than 3.5 *billion* years because there's always *all*. It has survived for more than 3.5 *billion* years because there's always *one*.

< ALL, > NONE

I have heard the phrase 'there's always one' a million times. When herding livestock you can almost guarantee that the last animal will charge in the opposite direction and after a while I realised that muttering farmers were right, because there is often, if not always, 'one'. It's a regular occurrence and I quickly began to wonder why? Why does one animal ever feel obliged to risk a lonesome charge in a random direction? Could it be a biological phenomenon akin to 'not putting all your eggs in one basket'? Could they be splitting risk and diversifying survival opportunities?

To be honest, I have no peer-reviewed idea, but if those animals were being sent to slaughter sticking together clearly *united* risk and *destroyed* survival opportunities.

Then I began to think about Life in general. I began to wonder if evolution was about generating maverick species, and renegade individuals, to split risk and diversify survival opportunities in a similar way.

Then I read *The Origin of Species*, and all reservations vanished. Under the sheer weight of Charles Darwin's evidence I was forced to accept that divergence is the key to surviving in a dynamic world with an unpredictable future. It's essential and 'there's always one' is the whole point. Charging blindly into unknown futures new variations split risk and diversify survival opportunities because if they don't they *unite* risk and *limit* survival opportunities. They have no idea what will happen;

they just branch off and hope for the best and by covering as many options as possible a few of them survive to continue the process.

If you want an example, consider this statement from the laboratories Biogéosciences (Université de Bourgogne / CNRS), Paléoenvironnements & Paléobiosphère (Université Claude Bernard / CNRS) and the Universities of Zurich and Lausanne (Switzerland) concerning a recent research paper in the journal Science (28/8/2009):

> 'The history of life on Earth has been punctuated by a number of mass extinctions, brief periods of extreme loss of biodiversity. These extinctions are followed by phases during which surviving species recover and diversify. The End-Permian extinction, which took place between the Permian and Triassic [about 250 million years ago], is the greatest mass extinction on record [and one which we will brielfy return to in the next chapter], resulting in the loss of 90% of existing species. The cephalopods [ammonites], which were abundant during the Permian, narrowly missed being eradicated during the extinction: only two or three species survived and a single species seems to have been the basis for the extraordinary diversification of the group after the extinction [they needed only one million years after the End-Permian extinction to diversify to the same levels as before].' [205]

i.e. there was 'one', not 'all'.

Incidentally, here are Charles Darwin's thoughts on the final extinction of ammonites in general (alongside the dinosaurs about 65 million years ago):

> 'In some cases, however, the extermination of whole groups of beings, as of ammonites towards the close of the secondary period, has been wonderfully sudden.'

Not exactly doom and gloom is it (I hope you haven't forgotten that the Charles Darwin Foundation is a wonderfully ironic member of the Alliance for Zero extinction).

Anyway, the point is the IUCN's Chief Scientist and the 'world's best natural history magazine' can say whatever they like, because they have clearly forgotten that evolution is about totally undefined variation, rather than precisely defined species, and that's a mistake that makes them look as deluded as they are morphologically racist. Despite their utopian fantasies, evolution isn't about 'alls', it's about 'ones'.

SPECIATION

Extinction *is* a biological reality of course. It's just not one that means an awful lot to races and individuals, especially after they're extinct. It's much more important to people who can project extrinsic values onto dead animals and such people aren't even happy to fight a constant battle. Indeed, the 'stewards' seem obsessed with finding the world's smallest species and that's a policy that can only create new extinction risks and more conservation. Admittedly, this obsession is based on the spurious research of speciation scientists, but rarer divisions are obviously exciting because the 'stewards' devour them with relish.

Well, most of them do anyway. Even some of the 'stewards' are a bit concerned about this trend. Here's regular BBC Wildlife Magazine contributor and general media 'steward' Mark Carwardine to explain:

> 'On paper, at least, species-splitting reduces population sizes and inflates the number of endangered species, painting an altogether bleaker picture. But whether this is good for conservation or could be seen as crying wolf – using science to make the situation appear worse than it was before – only time will tell.' [206]

I think I might take over from 'time' at this point and 'tell' Mark, and all other species-splitters, that species-splitting *is* 'crying wolf'. It's science for the sake of science and it has gone way beyond objective natural history. For instance, the 'discovery' of Malayan Tigers inside Indochinese Tigers in 2004 was based on the examination of 'three molecular markers: (1) 4.0 kb of mitochondrial DNA (mtDNA) sequence; (2) allele variation in the nuclear major histocompatibility complex class II DRB gene; and (3) composite nuclear microsatellite genotypes based on 30 loci'[207], which means very little to me and even less to a tiger I'm sure. The researchers claim to be trying to 'establish objective methods for subspecies recognition' and strictly speaking they did. They also had to resort to some pretty esoteric genetic wizardry to do it and the result is a distinctly defined *identical* subspecies with a small amount of sub-microscopically different…stuff.

Giraffes have received an equally supra-objective, sub-superficial taxonomic overhaul, as reported by the BBC Wildlife Magazine:

> 'As it turns out, the giraffe [which is actually 'the giraffes' remember] does indeed comprise more than one species. Modern genetic analysis has uncovered the complex evolutionary history of the world's tallest ungulate [extreme cosmetic value], and the findings will rock those involved in conservation.' [208]

I must just stop there quickly, because 'the findings will rock those involved in conservation' summarises this subject perfectly. Basically, everything has changed because of a DNA test even though nothing has actually changed.

Anyway:

> 'According to UCLA zoologist Robert Wayne and colleagues, there are a minimum of six giraffe species, at least one of which is on the verge of extinction.'

Remember, this new potential extinction is entirely based on a DNA test, and it can only be monitored by more DNA tests.

Spurious isn't it, but the 'stewards' can't seem to get enough of it. They're desperate for species to splinter and luckily there are plenty of new 'species' to choose from. Scientists have recently discovered a new species of bottlenose dolphin for example, based entirely on the following premise:

> 'There are no clear morphological characters that can be used to distinguish genera and a number of species in this sub-family [Delphinidae].' [209]

A mere lack of 'clear morphological differences' wasn't enough to stop these guys though, and having split the Common Bottlenose Dolphin into the Common Bottlenose Dolphin and the Indo-Pacific Bottlenose Dolphin in 1998, the race was on to further fragment the species. This was finally achieved in 2008 and here's how they did it:

> 'Genetic variation was estimated by calculating number of alleles, and expected (He) and observed (Ho) hetero-zygosities in Genepop 3.4 (Raymond and Rousset, 1995). Tests for Hardy-Weinberg equilibrium (HWE), using an exact test and based on 1000 iterations, and tests for linkage disequilibrium were also conducted in Genepop, with the significance level Bonferroni-corrected (Rice, 1989).'

Still with it? If you're finding it confusing, please remember that this is how species are defined these days:

> 'We used a Bayesian clustering method implemented in Structure 2.1 (Pritchard et al., 2000) to test the as-signment of individual samples to genetic clusters. The number of clusters (K) was inferred from the posterior probability distribution Pr (K/X) calculated

from the posterior probability of the data Log Pr (X/K). For this analysis we used a burn in period of 100,000 iterations, runs of 106, values of K between 1 and 4, series of 5 independent runs for each value of K, with the admixture and correlated frequency models.'

If you managed to read it all, you did better than I did. I just resorted to cutting and pasting in the end because it's clearly esoteric nonsense.

It *has* generated the southern Australian Bottlenose Dolphin however, and even though you will never be able to recognise it, that's another species to add to the diversity pile and, more importantly, the emotive, cash-generating potential extinction list. The dolphins themselves couldn't care less of course, but this isn't about dolphins, this is about conservation.

The 'stewards' have had even greater success with the Australian gecko genus Diplodactylus. A recent survey of 'mtDNA, allozyme and chromosomal variation'[210] has increased the number of species from 13 to 29, which is typical 'of many recent studies in demonstrating how morphological data [i.e. having a look at them] can be ambiguous and even misleading.' Apparently, 'this level of undescribed diversity has serious implications', because 'a major problem for biodiversity conservation and management is that a significant portion of species diversity remains undocumented (the 'taxonomic impediment')', but it might be worth exploring Darwin's views on the 'taxonomic impediment' at this stage:

> 'Some few naturalists maintain that animals never present varieties; but then these same naturalists rank the slightest difference as of specific value; and when the same identical form is met with in two distinct countries, or in two geological formations, they believe that two distinct species are hidden under the same dress. The term species thus comes to be a mere useless abstraction, implying and assuming a separate act of

creation. It is certain that many forms, considered by highly-competent judges to be varieties, resemble species so completely in character, that they have been thus ranked by other highly-competent judges. But to discuss whether they ought to be called species or varieties, before any definition of these terms has been generally accepted, is vainly to beat the air.'

I guess the 'taxonomic impediment' doesn't mean that much when you're well aware that variation is the whole point.

The 'stewards' have even split a species because of slightly different 'accents', and this time it's been done specifically by a conservation group who are desperate to find a restricted population that they can shower with their warm, moist 'love'. Here's the RSPB to explain:

'The Scottish Crossbill was originally thought to be a sub-species of the common crossbill. However, when it was discovered that the two birds nested in the same wood but did not interbreed, the Scottish crossbill was given full species status and became Britain's only endemic bird.' [211]

They do interbreed of course, as discovered by none other than the RSPB, in the same study:

'The study established that there was a small percentage (4.3%) of mixed species pairing in relation to bill size, but this was insufficient not to regard the Scottish crossbill as a species.'

'Insufficient' clearly depends on your perspective and, rather unsurprisingly, not everybody was convinced. Sensing a threat the RSPB decided to have a good rummage through crossbill morphology to see if they could find something more...real:

'There are no plumage differences between Scottish, common and parrot crossbills, and although the Scottish crossbill is intermediate in size between common and parrot crossbills, it overlaps in size with both.' [212]

By now they were really beginning to panic, as I'm sure you can imagine. Time to search the genome then:

'No differences were found among any of the three crossbill types – not even common and parrot crossbills, whose species status has never been questioned.' [213]

Crumbs. What on Earth should they do?

'Sonograms...' [214]

By gum, I think they've cracked it:

'Sonograms of the calls of crossbills of known bill size (and, therefore, putative identity) have shown that the three crossbill species have different calls.'

Finally the evidence they needed and the RSPB's senior species-splitter, Dr Ron Summers, was delighted:

'This research proves that the UK is lucky enough to have a unique bird species that occurs here and nowhere else – and this is our only one.' [215]

Outrageous lunacy isn't it, and hideously wrong on so many levels. For example, 'the UK is lucky enough to have a unique bird species' completely ignores the fact that Crossbills, of whatever slightly different type, belong to themselves, not countries. They especially don't belong to people who are desperate to divide and possess them that's for sure. Worst of all though is the plainly demented assertion that an accent justifies, no, 'proves' a new species. Imagine applying such crazed divisive policies to the human race. I have no idea how many languages there are, never mind accents, but that would rapidly divide us completely, if not totally. It's all madness and thus I have absolutely no doubt that species-splitting *is* 'crying wolf'. Although, and I'm a little bit embarrassed to have to admit this, I'm not exactly sure which subspecies of wolf.

Talking of 'the wolf', I have one more example of speciation. To be fair to the 'stewards', this is more a case of geographic isolation than genuine taxonomic speciation, but a subpopulation *has* formed and the result's the same. This time it's the wolves of Isla Royale.

I mention them specifically because their geographical isolation has led to pedigree-dog type inbreeding and genetic problems are the result, but I must first point out that Isle Royale is located in the northwest portion of Lake Superior (on the border between Canada and the USA), and that it has been home to a small wolf population since about 1950. This currently consists of about 24 wolves living in three packs, although these numbers do fluctuate from year to year.

Onto the problem then, as reported by The Wolves and Moose of Isla Royale (not literally, that's just the title of the attending research project):

'The wolf (Canis lupis) population on Isla Royale, a remote island in Lake Superior, North America, is extremely inbred. Nevertheless, the consequences of genetic deterioration have not been detected for this intensively studied population, until now.' [216]

Here we go then:

'We found that 58% of Isle Royale wolves exhibited some kind of congenital malformation in the lumbo-sacral region of the vertebral column and 33% exhibited a specific malformity, lumbosacral transitional vertebrae. By contrast, only 1% (1 of 99) of wolves sampled from two outbred, wolf populations exhibited this malformity.'

In case you missed it, 58% possess congenital deformities. I wonder if there's a possible management solution?

'The future occurrence of these malformities could possibly be reduced by genetically rescuing the wolf population.' [217]

A genetic rescue? That's a big step. I assume there's some legitimate responsibility that could justify such 'stewardly' intervention:

'Wolves…arrived by walking on an ice bridge from Canada.' [218]

Clearly not. The truth is these wolves made their own way to Isla Royale because 'there's always one' and because 'there's always one' is a fundamental part of 'branching out and seizing on many new places in the polity of nature'[Darwin]. They made a break for it which is beginning to fail. Not in an unusual way, but in the same way as trillions of blind excursions before it and yet the 'stewards' just can't stand by and let nature be nature. They have lost themselves in their own self-reverence again and are now considering a perfectly anthropomorphic 'genetic rescue'. I should point out that they do concede that 'the appropriateness of attempting a genetic rescue depends on technical matters and important, unresolved ethical issues…'[219], but this case isn't amazing because they're not sure about a genetic rescue; this case is amazing because it's being considered at all. It's pure conservation and I can clarify the 'ethics' in one sentence: managing nature is not observing nature. Beyond that they can do whatever they like.

It's also interesting to note that:

> 'The results are significant because many policy makers and stakeholders and some conservation professionals use examples like Isla Royale wolves to downplay the consequences of genetic deterioration.' [220]

But that's drifting slightly from the subject. The point is the Isla Royale wolves aren't just wolves anymore, they're Isla Royale wolves, and even though their only distinction is self-imposed geographical isolation (well, imposed by the original wolf pioneers anyway) that has generated a whole new extinction possibility and a whole new conservation priority. In the words of Kris Hundertmark, a University of Alaska Fairbanks wildlife geneticist at the Institute of Arctic Biology:

> 'When we give something its own name we're saying this is a unit of biodiversity that deserves to be conserved. If you name something that doesn't deserve a name, you're wasting resources that could be spent on worthwhile groups.' [221]

Which is yet more completely typical conservation madness. This time group value has nothing to do with the intrinsic mission of the individuals within the group, or even their superficial appearance. Instead it's entirely based on a name.

> All animals are equal but some animals can shove off because they don't have a name they couldn't care less about anyway.

Anyway, what seems to have been forgotten, again, is that every individual of every species faces its own extinction every day, and that includes you. You may disagree, but I'm sure most normal people wouldn't ignore their own extinction in favour of that faced by 'the human', so why do conservationists do it to species? Especially ones they keep dividing all the time. It's compulsive obsessive anthropocentric taxonomic melodrama. I guess individuals will eventually become their own species (thus reuniting death and extinction once again), but until that point how the hell is anybody supposed to know what they're looking at? Take Blue Whales for example. They have already been split into three different subspecies, but how do you know they won't be divided into Common Blue whales and 'Morphologically-Similar-but-Biochemically-different' Uncommon Blue whales; or Blue whales, Indigo whales and Azure whales?

The answer is: you don't know. With the 'stewards' chasing down spurious variations between alleles, mtDNA and major histocompatibility niblets anything can happen. It may seem unlikely now, but before DNA testing it was pretty unlikely that the African elephant would become the African Forest Elephant and the African Savannah Elephant[*]; or that the Goliath Grouper would become the Pacific Goliath Grouper and the Atlantic Goliath Grouper; or that the African Dwarf Crocodile would become the Congo Dwarf Crocodile, the Non-Congo (Ogooué Basin) Dwarf Crocodile and the Non-Named, Non-Congo, Non-Ogooué Basin Dwarf Crocodile; but that's exactly what's happened. Based on DNA evidence defined species are un-defining themselves all the time and Darwin's observations remain true:

> 'I was much struck how entirely vague and arbitrary is
> the distinction between species and varieties.'

How do you know some Snow Leopards aren't actually Sleet Leopards for example? Or that some Lemon Sharks aren't Lime Sharks? And how do you know that some Wandering Albatross' aren't actually Casually Meandering instead? And

---

[*] A third species of African Elephant has also been proposed. Using DNA evidence (again) some biologists reckon the West African populations are neither Forest or Savannah Elephants; they're Forest *and* Savannah Elephants instead.

what about the Good White Shark, is that a comedy metaphor, or an unrecognised species? With the speciators on the case how do you know anything for certain? You don't, and the concept of extinction is based around this uncertainty. It has existed for less than 400 years, and it's based on species that can't be definitively separated at the edges. They can be loosely separated into sets 'of individuals closely resembling each other'[Darwin], but it's all gone a bit far when individuals that look exactly the same can be separated into species which look…exactly the same.

And besides, the *only* reason they can be loosely separated at all is because the intermediate forms have been eliminated by natural selection. In the words of Professor Richard Dawkins:

'The illusion of a borderline is created by the fact that the evolutionary intermediates happen to be extinct.' [222]

If they had survived, current species would be linked by everything in between and it would be utterly impossible to tell where one species ends and another begins, but because they haven't species appear artificially distinct and the 'stewards' can thus lose themselves in a world of well-defined extinction that's being undefined and redefined all the time.

Perhaps the University of Gothenburg have summarised this speciation craze better than anybody:

'Animals that seem identical may be completely different…' [223]

Perfect.

REINCARNATION

At the most extreme end of subjective appearance worship is reincarnation, because that's about prejudice towards races that don't even exist anymore. Yet again, it has nothing to do with intrinsic value and everything to do with superficial appearance, but this time it's about animals that are already dead, like Galápagos Tortoises:

> 'Yale scientists report that genetic traces of extinct species of Galápagos tortoises exist in descendents now living in the wild, a finding that could spur breeding programs to restore the species.' [224]

Yale university will disagree of course, but does anybody really think that a 'genetic trace of an extinct species' cares about anything at all, never mind breeding programs to restore slightly different tortoises?

Yes, they do, but that's exactly why conservation is not natural history. Apparently, extinct races aren't just single units in time and space, they're also single units in time and space that are suffering because they no longer exist. It's philosophical madness and this kind of extrinsic nonsense isn't limited to Tortoises that make, in the word of Charles Darwin, 'excellent soup'. For instance, The Quagga Project of South Africa is trying to find an extinct subspecies of Zebra ('the Quagga') inside 'the Plains Zebra' because they think reincarnation will 'rectify a tragic mistake'[225], but 'tragic' for whom, because I'm quite sure dead Quagga's couldn't care less. I guess it's 'tragic' for people who really like Zebras to look a certain way and, sure enough:

> 'It is this marked difference in appearance between the extinct Quagga and its northernmost relative, the Grant or Boehm Zebra, which makes the re-breeding of the Quagga desirable.' [226]

And the marked difference worthy of genetic grave-digging is?

> 'It differed from other zebras mainly in having been striped on the head, neck and front portion of its body

only, and in having been brownish, rather than white, in its upper parts. The belly and legs were unstriped and whitish.' [227]

Basically, its backside wasn't stripy.

Some people believe that unstriped Zebra buttocks justify serious breeding programmes though and they're determined to succeed. Indeed, they would even consider cloning Quaggas if they could. Unfortunately however, they don't have enough genetic material and that is, as I'm sure you'll agree, a dreadful shame. It would be truly underwhelming to have a slightly different Zebra as soon as scientifically possible.

The slight differences were so enormous that nobody even noticed 'the Quagga's' original extinction in the first place:

'It was only realised years later that when the Quagga mare at the Artis Magistra zoo in Amsterdam died on the 12th August 1883, she was the last of her kind! The true Quagga vanished unnoticed.' [228]

Admittedly, humans were the primary cause of this extinction (because of the indiscriminate hunting of zebras in general), but extinction always has a cause and, given that just under every species that have ever existed is already extinct, extinction with a cause is not exactly unprecedented.

The Quagga Project disagrees however, and with their help 'the Quagga's [originally unnoticed] extinction may not be forever!' Some scientific, and point-missingly anal concerns have been raised of course:

'It has been argued that there might have been other non-morphological, genetically-coded features (such as habitat adaptations) unique to the Quagga and that therefore, any animal produced by a selective breeding programme would not be a genuine Quagga...[but] since there is no direct evidence for such characters and since it would be impossible now to demonstrate such characters were they to exist, the argument is spurious.' [229]

But, to be fair to the Quagga Project, the whole thing is spurious. 'The Quagga' is extinct, like millions of species and subspecies before it, and, regardless of what the 'stewards' think about the equality of selection, that's totally unfair in a totally normal way. And besides, 'the Quagga' that's extinct was 'the Quaggas' that faced inevitable personal extinctions anyway. Which applies equally to 'the Quagga' they're currently trying to find inside 'the Zebra' that couldn't care less.

A Quagga[230] *– Slightly different and thus terribly valuable.*

Wildlink International's desperate search for a subspecies of lion with a slightly darker mane ('the Barbary Lion') is another example. Actually it's even worse, because according to BBC Wildlife Magazine (again):

'Just because a lion has a large, dark-fringed mane doesn't mean it's a barbary. It's appearance can be influenced by external factors, such as the climate of the zoo in which it's held. As a result: the only sure way to identify a Barbary Lion is through DNA testing' [231]

Which means these people want to back-breed a lion sub-species just because its DNA is slightly different. You won't be able to tell by looking at it, you'll only be able to tell by the sign on the cage or a DNA test and that's a whole new level of taxonomic reincarnation madness. Can't they reincarnate something that's not about cosmetic appearance, like Bulldogs that can breathe for example, or Belgian Blue cows that can give birth, or Holstein dairy cows that can survive without drugs?

Of course not, they want Quaggas, and Barbary Lions, and slightly different tortoises, and Bucardos (Pyrenean Ibex), and Thylacines (Tasmanian Tigers), and Woolly Mammoths instead. *Woolly Mammoths for goodness sake.* Where the hell are they going to put Woolly Mammoths? Especially in the middle of a global warming period. Woolly Mammoths were an ice age species and these pumpkins want to reincarnate them now. Essentially, that's just creating them to kill them again.

Extantion (preventing or reversing extinction) is about justice though isn't it, not reality.

And saving the planet of course…

Chapter 8

PLANECIDE

'Nevertheless, so profound is our ignorance, and so high our presumption, that we marvel when we hear of the extinction of an organic being; and as we do not see the cause, we invoke cataclysms to desolate the world, or invent laws on the duration of the forms of life!'[Darwin]

'Cataclysms to desolate the world' don't come much bigger than the death of the entire planet, but that's the claim. Sometimes it's by insinuation of course, when it's packaged into claims about 'saving the planet', but either way, the death of an entire planet is a genuine concern for the 'stewards'.

Hasn't Chernobyl taught them anything though? For those who don't know, and according to the UN related International Atomic Energy Agency (IAEA), 'the accident at the Chernobyl nuclear power plant in 1986 was the most severe in the history of the nuclear power industry, causing a huge release of radionuclides over large areas of Belarus, Ukraine and the Russian Federation'[232]. It's the only nuclear disaster to have reached Level 7 (the maximum level) on the IAEA's International Nuclear Event Scale (INES) and the whole thing generated liberal speculation about nuclear wastelands and sterile holocaust deserts. It was described as an appalling ecological disaster with, in the words of Greenpeace, 'devastating consequences'[233], and the 'stewards' instantly and pessimistically assumed Homo sapiens had been powerful enough to destroy nature.

But they hadn't. They'd been powerful enough to claim a few early casualties, but, in the words of the IAEA:

'The present environmental conditions have had a positive impact on the biota in the Exclusion Zone. Indeed, the Exclusion Zone has paradoxically become a unique sanctuary for biodiversity.' [234]

It didn't even take that long for the wildlife to recover:

'By the next growing season after the accident, the population viability of plants and animals substantially recovered as a result of the combined effects of reproduction and immigration.' [235]

In fact, just four months after the accident 'fifty species of birds [and 'forty-five species of mammals']…all appeared normal in appearance and behaviour' and 'significant population increases of game mammals (wild boar, roe deer, elk, wolves, foxes, hares, beavers, etc.) and bird species (black grouse, ducks, etc.) were observed soon after the Chernobyl accident'[236]. Which means the most severe nuclear accident in human history has actually been a *good* thing for conservation, despite Greenpeace's typically pessimistic and melodramatic claims about 'devastating consequences'.

The 'stewards' are still desperately searching for reasons to condemn us of course. And it should be pointed out that some wildlife populations may have a higher incidence of deformities and a lower population density than surrounding areas. But the exclusion zone isn't a sterile desert and the Chernobyl disaster wasn't enough to kill nature.

The 'stewards' are also trying to deflect the credit as well. They're saying that Life is flourishing around Chernobyl because humans have been excluded for example, but that makes and breaks their case at the same time. There's no doubt the absence of humans has been involved, but there's also no doubt that change didn't 'invoke a cataclysm' and that it didn't 'invoke a cataclysm' specifically without the help of the 'stewards'. Basically, nature did a pretty good job all on its own.

Even *Homo sapiens* demonstrated how resilient nature is:

> 'Claims have been made that tens or even hundreds of thousands of persons have died as a result of the accident. These claims are highly exaggerated.' [237]

Actually, 'highly exaggerated' is probably an understatement, as the World Health Authority (WHO) will now explain:

> 'The estimates point to a total of several thousand deaths over the next 70 years, a number that will be indiscernible from the background of overall deaths in the large population group. The estimates do not substantiate earlier claims that tens or even hundreds of thousands of deaths will be caused by radiation exposures from the Chernobyl accident.' [238]

Please note, I'm not saying that people didn't die at all, because they did. Most of these deaths involved the emergency workers as well, which perfectly demonstrates why Life just isn't fair. Thyroid cancer has caused a few more deaths as well (less than 40 between 1992 and 2002), but basically humans have also proved nature's resilience.

Life has even invaded the reactor core itself. It has marched right into radiation central and lived to tell the tale. In fact, it's positively thriving. A mobile robot sent in to recce the place found black, melanin-rich fungus happily using ionising radiation to produce its own energy; i.e. while the prophets of doom wailed and moaned like spoilt cry babies Life marched straight up to the reactor core and started *eating* radiation. It laughed in the face of their pious sympathy and pessimistic cataclysms and yet, despite this amazing show of defiance, they still think nature is a baby lamb that needs to be wisely shepherded through change. And they still think humans can destroy the entire planet.

For the purposes of this argument I should point out that

I have assumed 'the planet' is a plural of 'lives', rather than a singular, omni-inclusive Gaia-type entity. Of course, if the 'stewards' are genuinely including the death of the non-living there's not an awful lot more to say, but I will give them slightly more credit than they have earned thus far, and assume their 'planet' and my 'nature' are the same thing.

Talking of Gaia, this is probably the right moment to introduce Professor James Lovelock. He will join Charles Darwin as an occasional contributor because some of his claims are as subjective as Darwin's are objective. That's not to suggest that Professor Lovelock isn't a remarkable scientist, because he is. He's responsible for recognising (and promoting in the face of extreme resistance) a single planetary ecosystem that regulates life-sustaining global conditions using a complex web of cross-linked inorganic and organic cycles. It's probably impossible to prove where universal physics ends and self-regulation begins of course, but the result is the same and he called this Earth system 'Gaia', after the Greek Goddess of the Earth. I have to admit, I don't know whether Gaia is a plural of lives or a super-massive 'one', but that's kind of the point, because neither does Professor Lovelock. In fact, it actually appears to be both:

> 'If Gaia does exist, then we may find ourselves and all other living things to be parts and partners of a vast being who in her entirety has the power to maintain our planet as a fit and comfortable habitat for life.' [239]

Inconsistent respect for the individual is just part of his confusion however, because his emotional interpretation of his own brilliant science has led him to all sorts of other conclusions that drip with subjectivity. Here are his thoughts on the familiar topic of whaling for example (remember, while whales are superficially extreme, they're no more focused on survival than Cane Toads or Grey Squirrels):

> 'It is bad enough to cull or farm the whale so as to provide a constant supply of those products which whale-hunting nations claim are needed by their backward and primitive industries [easy tiger]. If we hunt them heedlessly to extinction it must surely be a form of genocide.' [240]

Yes, you read it correctly. Killing whales is like genocide which, considering that dead whales are just dead, and his own assertion (in the same book) that 'large plants and animals are relatively unimportant'[241], is some pretty extreme extrinsic subjectivity, and there's plenty more:

> 'Darwin once described the evolutionary process as 'clumsy, wasteful, blundering, low and horribly cruel'[*]. But surely not as cruel, or as culpable, as we have been and still are to the rest of life on Earth; especially since so many other innocent organisms will share our fate.' [242]

'Innocent'? What the hell has 'innocence' got to do with evolution? Last time I mentioned 'innocence' it involved a seal fanatic who had turned her inter-species racism into a competition to win a holiday. This time it's a distinguished Earth scientist who is supposed to understand the difference between subjective and objective, and who also asserts that:

> '…it is so often ignored or deliberately forgotten that the unending death-roll of all creatures, including ourselves, is the essential complement to the unceasing renewal of life.' [243]

Do you think he ignored it, or just deliberately forgot?

---

[*] James Lovelock is referring to the following sentence from a letter Charles Darwin wrote to Mr. J. D. Hooker on the 13th July 1856: 'What a book a devil's chaplain might write on the clumsy, wasteful, blundering, low and horribly cruel works of nature!'

Here's another one:

> 'If we fail to curb global heating, the planet could mas-
> sively and cruelly cull us, in the same way that we have
> eliminated so many species by changing their environ-
> ment into one where survival is difficult.' [244]

We will come to 'global heating' in earnest later, and indeed human extinction. For now, the first thing you should notice is that being culled is only cruel if you're a species, and only if it's done by global heating to us, or by us to others. This is clearly nonsense. Firstly because you can't be cruel to a dead species, secondly because species aren't real units of selection, thirdly because *individuals* face culling every day and fourthly because being 'cruelly culled' is an essential part of the whole process. Remember, species don't evolve themselves in the right direction, they're evolved in as many directions as possible, and the *right* direction picks them.

The right direction changes all the time as well, which is why moral judgements about the iniquity of change are so pointless. In fact, invoking guilt by suggesting that environmental change involves 'cruelty' is subjective madness, especially when you find out that Professor Lovelock doesn't think we should feel guilty about anything anyway:

> 'Like the photosynthesisers, we could not have avoided
> reaching our current overpopulated and unsustainable
> state. We are what we are and there is little we could
> have done to avoid what now seem adverse changes;
> we should not feel guilty about it.' [245]

Although his forgiveness is not always obvious:

> 'We became the Earth's infection a long and uncertain
> time ago' [246] and 'like a drunkard driving a tank we
> have accidentally trashed our world.' [247]

Now, I'm not sure about you, but being described as a drunk infection that has accidentally trashed our world hasn't always filled me with guilt-free satisfaction. It doesn't have any effect now of course, but that's only because I have started seeing this kind of inconsistent rhetoric everywhere, especially in Gaian philosophy. Professor Lovelock doesn't even agree with the title of his own books sometimes. How does *The Vanishing Face of Gaia* (2009) fit with 'what people mean by the plea ['save the planet'] is 'save the planet as we know it''[248] for instance? I assume that what he means by 'the vanishing face of Gaia' is: the vanishing face of Gaia as we know it, but I'm not entirely sure, especially in light of other sentences from the same book:

> 'If we used these [Arctic fossil fuel reserves] as we use them now, we might become our own executioners and cause the death of Gaia as well.' [249]

I guess it means the death of Gaia's vanishing face as we know it, although there are more that pull the other way as well of course:

> 'The real Earth does not need saving. It can, will and always has saved itself.' [250]

Which are supported by others from a previous book (*The Revenge of Gaia*):

> '…there are no grounds for thinking that what we are doing will destroy Gaia.' [251]

Which are turned inside out (again) by others from the same previous book:

> 'Now humanity and the Earth face a deadly peril, with little time left to escape.' [252]

Which contradicts yet more from the original book (*Gaia – A new look at life on Earth*):

> 'It is now generally accepted that man's industrial activities are fouling the nest and pose a threat to the total life of the planet which grows more ominous every year. Here, however, I part company with conventional thought. It may be that the white-hot rash of our technology will in the end prove destructive and painful for our own species, but the evidence for accepting that industrial activities…may endanger the life of Gaia as a whole, is very weak indeed.' [253]

Hard to tell what's 'vanishing' and what's not isn't it. I guess the summary is that Gaia is going to live or die, possibly at the same time.

Inconsistency is not my main reason for introducing Professor Lovelock though. My main reason is the obvious similarity between 'Gaia's vanishing face' and conservationists 'death of the planet'. Indeed, the former may well have inspired the latter by linking everything together and helping conservationists decide that species and planets are the units of natural selection, rather than individuals and genes. To be fair to Professor Lovelock, he is now trying to distance himself from the unrealistic world of mainstream conservation. Unfortunately however, the unrealistic world of mainstream conservation isn't listening anymore. It has seized his baton and he only really has himself to blame, because books with diametrically opposed conclusions, like 'we might…cause the death of Gaia'[254] and 'there are no grounds for thinking that what we are doing will destroy Gaia'[255], just give people the opportunity to cherry pick the conclusions they like while ignoring the ones they don't.

Professor James Lovelock will represent two aspects of the following debate then. Firstly he will join Darwin as a spokesperson for objective evidence, and secondly he will represent the subjective doom and gloom rhetoric that typifies his unwanted flock of 'stewards'. Sometimes it will be difficult to tell the two apart, but that's kind of the point, because if this great man can tie himself in anthropocentric knots it's easier to understand how the 'stewards' have done it.

One more Lovelock nugget of pessimism before we continue:

> 'If we think of Gaia as an old lady still quite vigorous but nowhere near as strong as the young planet that carried our microbiological ancestors, it should make us realise more seriously the danger that we are to her continued healthy existence.' [256]

How patronising is that? I wonder if Professor Lovelock would be happy explaining how old and weak Life is to an angry Cape Buffalo, or a hungry Polar Bear, or a charging Lion. It's insulting, but, and lets be perfectly clear about it, this is *not* objective natural history. This is a 'cataclysm to desolate the world'[Darwin] and it specifically ignores the greatest experiment of all time. An experiment that has been running for more than three and a half thousand million years. An experiment called 'Life'.

> 3.5 BILLION YEARS

Life is definitely facing a challenge today, but, amongst the wails and ululations of the 'cataclysm invokers', it's often conveniently forgotten, or genuinely not known, that Life has *never* stopped facing a challenge. As discussed in Chapter Two, Planet Earth is not a static playground full of baby rabbits, it's a threatening environment full of unpredictable challenges instead. It's constantly changing and yet Life is still here. Despite 3.5 billion years of planetary wobbling (Milankovitch cycles) and wandering continents and a whole host of other threats, Life has endured and that simply can't be ignored.

Well, not if you respect evidence anyway. For those who can't nature will continue to consist of sacred species and peaceful seals, but for those who can, the great survival experiment has involved all sorts of glorious triumphs. One of the largest involved a pollution event that makes anthropogenic pollution look genuinely pathetic. According to Professor Lovelock it was:

'The most critical period of all in the history of life on Earth.' [257]

He doesn't say whether it was 'cruel' or not of course, but he does reveal the rather surprising cause:

'Oxygen…'

Oxygen?
Yes. Oxygen:

'Oxygen gas in the air of an anaerobic world must have been the worst atmospheric pollution incident that this planet has ever known.'

What you have to bear in mind is the complete lack of oxygen in the primordial atmosphere. It was totally anoxic (without oxygen) and the early biosphere was completely anaerobic (living without oxygen) as a result and that made oxygen a deadly poison and the evolution of photosynthesis a deadly threat.

To be fair, oxidation reactions consumed oxygen as fast as the early photosynthesisers could produce it initially. This bought the resident anaerobes a little more time (approximately 300 million years), but eventually production overwhelmed oxidation and atmospheric oxygen levels began to rise. After 2 billion years of happy anoxia the atmosphere became oxic (not a real word) and that must have reduced

Archean (2.5 - 3.8 billion years ago) and early Proterozoic (2.3 – 2.5 billion years ago) anaerobic biodiversity dramatically. It represented a great change in the conditions of life. Anaerobic life was pushed into the deep recesses of the biosphere and aerobic life took over. Which was quite fortunate really, because oxygen levels continued to rise, from 0%, through a high of 30-35% in the Carboniferous period (300-360 Mya) and on to today's level of about 21%. That's a rise of 210,000 ppm to now and a rise of 300,000 – 350,000 ppm at its height.

It's hard to know whether the unicellular cyanobacteria responsible felt as guilty then as humans do now, but it did happen, Life was responsible, and anthropogenic pollution is pathetic in comparison (humans have currently added 100 ppm of CO_2).

Oxygen isn't considered poisonous anymore of course, but only because Life has adapted. Marginalised anaerobic communities may disagree, but in retrospect the 'Oxygen Catastrophe' was a critical step in Life's evolutionary history. It unleashed multicellular evolution and the foundations of complex life, right up to and including *Homo sapiens*. One cells poison is another's evolution juice I guess, but either way Life survived a primordial weapon of mass destruction that still represents the single greatest act of biogenic pollution in the history of Life on Earth.

Life proved its resilience again during 'Snowball Earth'. In fact, it may have proved itself again during *several* 'Snowball Earths'. I should point out that 'disagreement remains regarding the precise influence of glaciation recorded in sedimentary rocks deposited during the Neoproterozoic [1000-542 million years ago]'[258], but given that 'many workers have presented a sufficient inventory of sedimentary characteristics to support the contention of a profound icehouse in that era', including evidence suggesting that 'more than 600 million years ago, ice occupied tropical latitudes', I'm still happy to say that Earth was a big planetary 'snowball' and Life did survive.

For those who missed it, that's a 'profound icehouse' during which 'ice occupied tropical latitudes'. Honestly, how much more proof does anybody need?

Planet loads unfortunately, and the reasons why are painfully predictable. Most obvious will be the speed of change defence, and it will go something like this:

> But the Oxygen Catastrophe and Snowball Earth developed over millions of years, and anthropogenic planecide is happening in just a few centuries.

i.e. change can change things but only while the rate of change doesn't change, which is a quantitative difference that only matters to humans and a judgemental luxury that Life simply can't afford. And besides, how quickly do they think conditions changed when a multi-billion tonne asteroid smashed into the planet? Please note, I'm not saying the Chicxulub asteroid caused the extinction of the dinosaurs. I'm just saying that when an object bigger than Mount Everest ploughs into a planet at 100,000 km/hour, and with the force of just over 6.5 BILLION 'Little Boy' Hiroshima bombs[*], it will probably change the conditions of life pretty quickly. Especially when you consider the multi-thousand feet hyper-mega-tsunamis, and the forest-igniting super-heated rock vapour, and the earthquake and volcanic eruption triggering terrestrial shock waves, and the acid rain, and the impact winter[†], and the post-impact-winter impact summer etc. In fact, it seems perfectly obvious to me that the conditions must have changed in just a few hours and yet the 'stewards' still manage to claim that humans are planecidal.

---

[*] Using estimates of 400 zettajoules (400×10^{21} joules) of energy for the Chicxulub impact, and 63 terajoules (63×10^{12}) for the 'Little Boy'.
[†] An impact winter would have been the direct result of solar interference by airborne dust, ash and dissolved sulphur dioxide (which reflects sunlight before it reaches the ground).

Change can change things but only while the rate of change doesn't change.

Whether the Chicxulub asteroid killed the dinosaurs or not is a different matter. All we really know is that something killed the dinosaurs at about the same time as a 400 zetajoule mega-impact, and, if it wasn't that, I dread to think what it was.

I will force myself to ignore the dread for now though, because the concurrent eruption of the Deccan Traps is another dinocidal possibility that further illustrates nature's collective resilience. This time it had to survive a massive flood basalt event (basalt rock flood) that covered an area of 1-1.5 million km2 (in western India) in thousands of metres of rock. The rock itself was a local event of course, but the associated volcanic gases would have driven a volcanic climate event that *wouldn't* have been local. It would have been global instead and it would have presented another severe challenge that Life *did* survive.

Now, I'm not saying there wasn't a mass extinction 65 million years ago, because there was. I'm just saying that Life didn't get all tearful and dramatic by invoking 'cataclysms to desolate the world'[Darwin], it just trimmed the fat and used blind, unconscious evolutionary variety to find a path through the changing conditions.

The Deccan Traps flood basalt event was proceeded, 185 million years earlier (i.e. 250 million years before now), by the Siberian Traps flood basalt event, and that was even bigger. It lasted for thousands, perhaps even millions of years, and occurred at the same time as, according to the British Natural History Museum, 'the most severe mass extinction of all time'[259]. An extinction event known as 'The Great Dying'. In case you're wondering, it was called 'the Great Dying' because 90% of all recorded marine species and 70 percent of all recorded terrestrial vertebrate species…died. It's also called the Permian-Triassic extinction event and it's hard to imagine how it's not linked to the production of vast quantities of basaltic lava and associated volcanic gases that were occurring in Siberia at the same time. We will never be able to prove the link of course, because we're not there, but this mega-eruption did happen and Life did survive.

I could go on and on.

These aren't just interesting curiosities from history though; these are catastrophic events that would make humans in general, and the 'stewards' in particular, cry like little girls. They are genuine survival triumphs that reveal nature in all its resilient glory, and conservation in all its sniveling arrogance, and there are thousands, perhaps millions more. Some bigger, some smaller, but all challenging, and every single one has been ignored by the 'stewards'. How else could the World Wide Fund for Nature (WWF) produce this kind of nature management garbage:

> 'The Earth is at a critical point where the decisions and actions taken by one species – ours – will determine the future of all life.' [260]

The future of all life? Determined by one species? Are they mad?
They continue:

> 'We seek to save a planet, a world of life…We seek to be the voice for those creatures who have no voice.' [261]

I have absolutely no doubt that they don't mean Cane Toads. In fact, I have absolutely no idea which creatures they *do* mean. Clearly they're not talking about the trillions x $10^{\text{trillions}}$ of creatures that are already dead, or the trillions that die every day and that only really leaves the creatures they have artificially merged into species units (all = one).

Dreadful isn't it, but there's more:

> 'In order to survive and prosper, we must urgently change our course towards a healthy planet where

people and nature thrive in a stable environment, now and for generations to come.' [262]

I'm sorry, did they really just suggest that humans can stabilise the environment 'now and for generations to come'? Yes. They did. The absolute worst is yet to come though:

'We seek to instil in people everywhere a discriminating, yet unabashed, reverence for nature.' [263]

Reverence for nature? How the hell is anybody supposed to feel a 'reverence for nature' if 'nature' can't even look after itself anymore? I wouldn't mind if this was some fringe cult with no followers, but this is a mainstream environmental charity that's claiming to represent objective natural history, and they're not alone:

'The Wildlife Conservation Society [WCS] saves wildlife [including all individuals?] and wild places world-wide…WCS is committed to this mission because it is essential to the integrity of life on Earth.' [264]

Essential to the integrity of life on Earth? Does the Wildlife Conservation Society really believe that it and its mission are 'essential to the integrity of life on Earth'?

'Being stewards of the Earth at this point in history is no small task.' [265]

Yes, they do. Honestly, what's wrong with these people?

The Wildlife Conservation Society is the conservation charity that offers the opportunity to 'see tigers do puzzles' remember. It shouldn't be confused with the Wildlife Conservation Network, or the Wildlife Conservation Union, or any of the Wildlife Conservation Trusts, or any of the Wildlife Conservation funds, or any of the Wildlife Conservation groups, or any of the Wildlife Conservation foundations, or any of the Wildlife Conservation associations. Honestly, it's like something from *Monty Python's Life of Brian* isn't it[*], and I haven't mentioned Conservation International yet either:

'We are optimistic that life on Earth can be preserved.' [266]

At least they're optimistic I guess, but given 3.5 billion years of magnificent resilience, why is there even any doubt?

Who knows, but there is, and it's not just limited to the conservation charities either. Here's an extract from Kofi Annan's (former United Nations (UN) Secretary-General) message to 'celebrate' World Environment Day in 2001 for example:

'All of us have to share the Earth's fragile ecosystems and precious resources, and each of us has a role to play in preserving them. If we are to go on living together on this earth, we must all be responsible for it...no one is immune from the consequences of climate change, the destruction of biodiversity, or other grave threats to the environment. As we embark on a new century, let us resolve to adopt a way of life that can be sustained right through it. Let us be good stewards of the Earth we inherited from our parents. And let us preserve it for our children, and their children after them.' [267]

Likewise, 48 Nobel Prize winning scientists, including nineteen for physics, seventeen for medicine and twelve for chemistry, have decided that we (President Bush and his administration in particular, but humans in general) are: 'threatening the Earth's future'[268]. While the chairman of the 'greatest showcase

---

[*] 'Are you the Judean People's Front?' etc. etc.

in the world for natural history photographers [self-certified]'[*], Mark Carwardine, thinks we all need to make sacrifices to 'save the Earth'[269]. And the editor of the 'world's best natural history magazine [self-certified]' (BBC Wildlife Magazine), Sophie Stafford, is *always* highlighting 'the threats our fragile planet faces'[270].

These are the political, scientific and natural history elite remember, deciding that 'our fragile planet' needs assistance just because it's cutting loose sensitive species and ecosystems, like it has done countless times before. Even the British Natural History Museum agrees. *The British Natural History Museum for goodness sake.* Home of objective biological research and Charles Darwin's formidable legacy. Temple of nature and epicentre of rational life sciences. Even that *great* institute of impartial natural history has lost itself in human self-reverence:

> 'The future of natural diversity and the essential con-
> nections that allow life to flourish here on Earth rests
> in our hands.' [271]

One more time:

> 'The future of natural diversity and the essential con-
> nections that allow life to flourish here on Earth rests
> in our hands.'

In case you have missed the significance of that statement, that's the British Natural History Museum claiming that the future of *all* Life on Earth rests in the hands of *Homo sapiens*. What the hell is going on? The galleries and website[†] of that

---

[*] The Wildlife Photographer of the Year competition is the 'greatest showcase in the world for natural history photographers' according to an article Mr. Carwardine wrote in *BBC Wildlife Magazine* anyway (February 2008). To be fair, it is pretty good.
[†] www.nhm.ac.uk

great museum pay reverent and objective homage to 3.5 *billion* years of Life's incredible resilience, and then plummet into the murky depths of mawkish sentimentality as soon as modern humans come along ($\approx$ 200,000 years ago). They make claims like:

> 'Since the beginning of time, the world has undergone radical changes.' [272]

> 'Slow or fast, change is inevitable.' [273]

> 'Sometimes change comes with terrifying speed.' [274]

> 'The Earth is a dynamic planet and restlessness is part of its nature.' [275]

> 'Nothing lives forever neither body nor solid rock.' [276]

> 'Life is part of the restless surface.' [277]

> 'Nothing is certain except change.' [278]

And then decide to 'invoke cataclysms to desolate the world' [Darwin] just because humans are involved:

> 'Human action may destroy all life on Earth unless we learn to sustain our environment.' [279]

Destroy all life on Earth? How arrogant are they?
The museum even manages to claim that:

> 'Climate change, and other crises such as habitat destrution, species loss and pollution, have created unprecedented pressures on the planet's biodiversity.' [280]

Unprecedented? Have they forgotten that 'Since the beginning of time, the world has undergone radical changes', and 'slow or fast, change is inevitable' and 'nothing is certain except change'. Have they forgotten that Life has survived toxic atmospheres, meteor impacts, volcanic winters, extensive glaciation and a whole host of other extreme events that would make humanity tremble like an autumn leaf. Life is even sitting quite happily in the Chernobyl reactor core *eating* radiation, but still they claim that 'human action may destroy all life on Earth'. Can there really be a better example of subjective self-veneration?

Luckily, 'our fragile planet' has a few 'voices' to speak on its behalf these days. There's the WWF of course, and all the Wildlife Conservation…charities, but there are plenty more besides. Indeed, being a planetary 'voice' is the entire point of Greenpeace:

'Greenpeace exists because this fragile Earth deserves a voice.' [281]

And the RSPB actually claims to be 'nature's voice'[282]. Although so does the Natural Resources Defence Council (NRDC)[283] and they can't both be right.

Friends of the Earth exist because the Earth needs… 'friends' (obviously) and when you combine the lot it's hard to imagine how 'this fragile Earth' ever managed to survive alone isn't it. Greenpeace and Friends of the Earth didn't exist before 1971 for example, but how on Earth did 'this fragile Earth' survive for BILLIONS of years without Greenpeace and Friends of the Earth? I know 38 years represents just one ninety-two-millionth (1/92,105,263 to be precise (ish)) of the total struggle, but thank goodness they showed up eventually.

Honestly, all these 'friends' and 'voices'. The system has worked *with* change for *billions* of years, but suddenly, and based entirely on the fact that they don't like it, the 'stewards' have decided to fight *against* change. They've decided that conservation ('survival of everything', except cane toads of course) is better than evolution ('survival of the fittest') and it's just embarrassing. It's arrogant and condescending and it's all based on this mythical 'fragile Earth'. But this Earth is *not* fragile, and it's still showing us even today. Cane Toads are an obvious example, but there are many more. There are pathogenic bacteria that have evolved resistance to antibiotics, and malaria parasites that have evolved resistance to anti-malarial drugs, and the Ug99 'stem-rust' fungus that has evolved resistance to anti-rust genes (used to protect nearly all the world's wheat), and rats that have evolved resistance to warfarin, and many more besides.

Humans may not like them, but we're talking about objective natural history, not anthropocentric self-interest, and, as Professor Lovelock has objectively recognised, 'the Earth has not evolved solely for our benefit.'[284] We can't kill everything, no matter how hard we try and the emergence of resistance to Dichloro-Diphenyl-Trichloroethane (DDT) is a prime example.

For those who don't know, DDT was used extensively in the 50's and 60's to kill important pathogen vectors like mosquitoes, lice and other 'verminous' insects (i.e. insects that humans don't like), but it also caused an obscene amount of collateral damage. An obscene amount of collateral damage that helped inspired the American biologist Rachel Carson to write *Silent Spring* (1962), which was the book that effectively launched mainstream conservationism.

As hard as we tried though, and despite its devastating potency, DDT couldn't beat nature and insect resistance is now widespread. A lot of birds did die of course, including the songbird that inspired the title, but apart from the fact that every single one of those birds would be dead by now anyway, the disease vectors and crop 'pests' that humans tried to destroy have evolved to cope and the very battle that launched the 'fragile Earth' hypothesis has now proved it's not true.

Lives are fragile, Life is not.

Incidentally, it seems perfectly obvious to me that if the

'stewards' really want to protect the environment malaria *et al* are the perfect solution. Education and tourism and other fluffy conservation strategies are all well and good, but if you really want to prevent damage the first thing you should do is turn humanity into an unfavoured race. And the absolute last thing you should do is, in the words of the British Natural History Museum, 'control the species that pass the disease to us'[285]. Apart from the anthropocentric racism against the *Anopheles* group of mosquitoes (who are just trying to survive remember), the *Anopheles* group of mosquitoes, and the malaria parasites they transmit, are nature's own 'stewards' and eliminating them just makes the job that much harder.

For all those who think I can only make such a statement because I haven't had malaria, I have had malaria and it's not pleasant, but that's exactly why it's capable of reducing human impact. And besides, I'm *not* advocating such a policy. I'm just saying that conservation would be much easier if conservation areas were full of life-threatening diseases.

The WWF disagrees of course:

'Clearly, there is no room for slippage in the fight against malaria.'[286]

But you must remember that malaria is a rival 'steward' that might actually achieve something.

To be fair to the WWF, I'm being terribly unfair. The conclusion remains the same though, because malaria would protect the environment better than good intentions and the invocation of cataclysms. Indeed, Professor Lovelock has gone even further down this line of thinking. He has suggested that we should specifically consider polluting important areas with nuclear waste to exclude humanity:

'The preference of wildlife for nuclear-waste sites [like Chernobyl] suggests that the best sites for its disposal are the tropical forests and other habitats in need of a reliable guardian against their destruction by hungry farmers and developers.' [287]

But that's a good plan that will inevitably fall on deaf ears.

Anyway, back to the mongering of doom and having considered at least 3.5 *billion* years of nature's amazing resilience it seems quite clear to me that Life will not end, even if some lives will. Perhaps the IUCN can summarise the pervasive invocation of cataclysms better than I can though:

'Fear can be a great motivator, and it is one that environmentalists have long cherished.' [288]

Can you believe that? What a perfect summary.

Maybe I'm being a bit unfair though. Maybe there is actually a genuine difference between this 'cataclysm' and all those that have gone before. Perhaps Professor Lovelock can explain:

'Change is a normal part of geological history…What is unusual about the coming crisis is that we are the cause of it.' [289]

I guess not. I guess that in light of all the evidence, and according to the internationally acclaimed Earth scientist responsible for one of the most important environmentally apocalyptic books in history, the most unusual thing about the coming crisis is…the cause.

Does that mean the climate has changed before then?

GLOBAL WARMING

Before charging objectively into the reality of climate change, I would like to say that global warming *is* real and that humanity *is* involved. There can be little doubt left and anybody who can turn 'little doubt' into complete denial will enjoy future discussions about belief without evidence and faith without fact (see Chapter Ten). For everybody else the modern climate *is* changing and CO_2 is heavily involved. That's not the issue. The real issue is whether the climate *has* changed before and whether a warm Earth is a dead Earth.

I would also like to say that I have chosen to focus on global warming because it's the flagship cataclysm that defines modern conservation. There are other issues of course, but global warming is linked to almost all of them in some way or another and it therefore offers a convenient umbrella threat with which to focus the discussion.

I would finally like to say that global warming will probably be a horrific disaster for *Homo sapiens* (and many others I'm sure). We have evolved to cope with different conditions and we will probably prove to be pretty sensitive to the rapid changes currently occurring. We will discuss human sustainability in the next chapter though so I will skip it for now and simply remind you that anthropocentric natural history is not objective natural history and that 'the Earth has not evolved solely for our benefit'.

Let's crack on then and at the centre of the global warming debate is carbon dioxide (CO_2). There are ancillary gases as well of course, but, because of its potent ability to insulate the planet, CO_2 is the main offender and the 'stewards' are quite certain that atmospheric levels need to be lower. They do accept that it's a normal by-product of respiration of course, and that it's a vital part of the biochemical process that supports most life on Earth (photosynthesis[*]), but they don't accept that levels can go much higher without generating the end of the world. The trouble is, prior to the industrial revolution atmospheric CO_2 concentrations had never been any lower.

---

[*] Photosynthesis is the biological process that turns carbon dioxide, water and sunlight into sugar and oxygen ($6CO_2 + 6H_2O \rightarrow C_6H_{12}O_6 + 6O_2$) and thus sunlight into energy that's available to the biosphere.

Well, that's not strictly true; they had, but only when there was no atmosphere at all. When the early atmosphere established itself over 4 billion years ago CO_2 concentrations were somewhere 250,000 parts per million (ppm) and 800,000 ppm and basically they have been dropping ever since. For instance, when photosynthesis evolved, during the Archean eon CO_2 concentrations were about 30,000 ppm; and when evolution went into overdrive, during the Cambrian explosion, CO2 concentrations were about 5,000 ppm; and just before the industrial revolution CO_2 concentrations were about 280 ppm. If we accept 250,000 ppm as the starting figure then, that's a 99.888% drop (Figure 1).

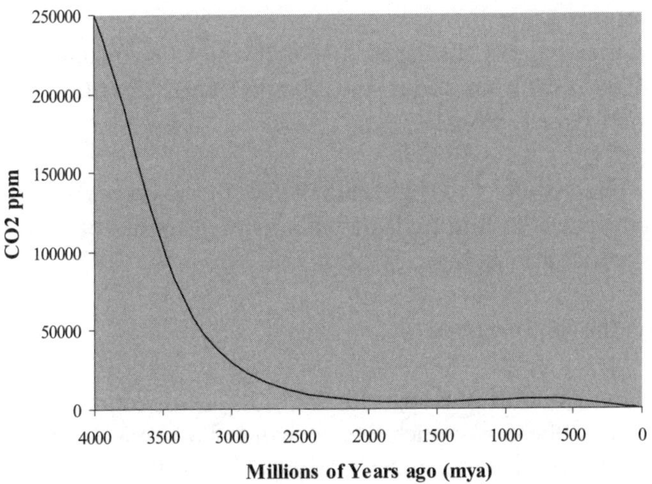

Figure 1 – *CO_2 over the last 4 billion years.*[*]

------

[*] Please note, this is a very simple version of the general trend rather than a comprehensive review of all current data. That would have a lot more wiggles and spikes. Please also note that to maintain consistency, and unless otherwise stated, subsequent graphs will use exactly the same scales. Please finally note that life probably began at some point between 3.5 and 4 billion years ago.

In case you're interested, these figures have been generated using isotope ratios in ancient chemical markers, fossilised plant data (tree rings, pollen morphology and stomatal density[*]), sedimentary rock compositions, geochemical computer models and a whole host of other clever proxy measurements and the implications are widely accepted by mainstream climate scientists. For instance, in its third assessment report the Intergovernmental Panel on Climate Change (IPCC) freely discussed Palaeo CO_2 and natural changes in the carbon cycle based on these results:

> 'There is evidence for very high concentrations (>3000 ppm) between 600 and 400 Myr BP [million years before present] and between 200 and 150 Myr BP'[290]

And they happily displayed data showing CO_2 concentrations of 4500-6000 ppm about 450 Myr BP. In their intergovern-mental-panelly words:

> 'The net effect of slight imbalances in the carbon cycle over tens to hundreds of millions of years has been to reduce atmospheric CO_2.'[291]

Even the WWF agrees:

> 'Four billion years ago its [CO_2] concentration in the atmosphere was much higher than today (80% [800,000 ppm] compared to today's 0.03% [300ppm]).'[292]

Al Gore doesn't of course:

---

[*] The stomatal density of modern and ancient leaves is regulated by the need to absorb CO_2 and the need to reduce water loss. Higher densities mean low CO2 has forced plants to risk increased water loss, and lower densities mean high CO2 has allowed plants to limit water loss.

'At no point in the last 650,000 years before the pre-industrial era did the CO_2 concentration go above 300 parts per million.' [293]

But Mr. Gore has failed to realise, or failed to reveal, that 650,000 years represents just 0.019% of the history of Life on Earth, which is a pretty pathetic sample on which to base judgement day (please see figure 2).

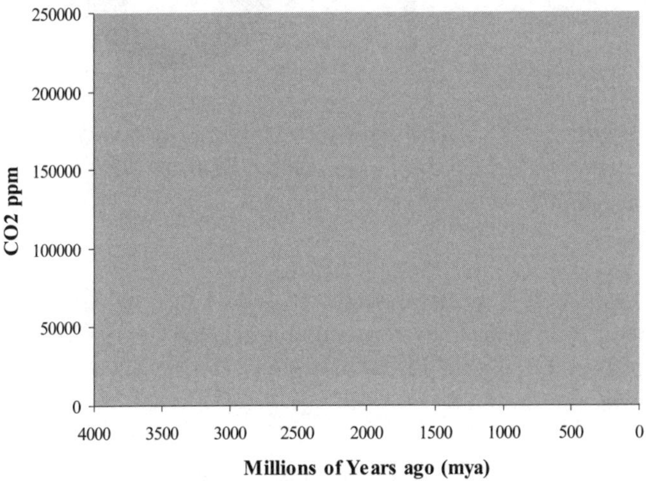

Figure 2 - CO_2 over the last 650,000 years. You can't see Al Gore's reference range because, as previously mentioned, the scales on this graph are exactly the same as before. I don't want you to think I'm being unfair though, so to prove that I did include his area of interest, I have magnified the bottom right hand corner of this graph by over 1000% (Figure 3 overleaf).

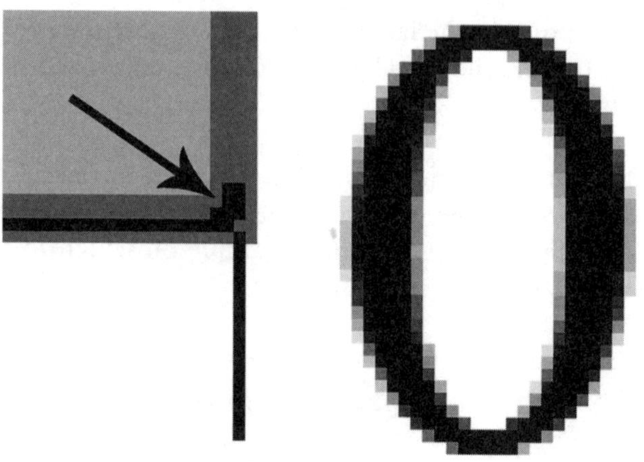

Figure 3 - *CO$_2$ over the last 650,000 years The changes that matter to Mr. Gore are indicated by the arrow. The '0' was originally below the vertical line.*

Al Gore is not alone though. In fact, Captain Apocalypse himself (evangelical environmental journalist George Monbiot) also uses 0.019% of Life's history to define 100% of Life's future:

> 'the levels of carbon dioxide and methane in the atmosphere…are now higher than they have been for 650,000 years.' [294]

Indeed, George Monbiot has even written a book with the title *Heat: How we can stop the planet burning*, as if the planet is literally going to burst into flames.

Incidentally, the front cover of this book proudly announces 'George Monbiot's myth-busting', but given the information that follows, it's hard to imagine a bigger myth than his totally unprecedented planetary fireball.

Anyway, apart from Al Gore and George Monbiot, a lot of climate specialists agree that atmospheric CO$_2$ has been *falling* for billions of years. This is particularly well recorded for the Phanerozoic climate (0-542 Myr BP). That has been reconstructed using a number of different proxies, including stomatal densities, paleosols (fossilised soils) and geochemical modeling, and the results were used in the IPCC's fourth report (2007) to show that CO2 concentrations have dropped from about 5000 ppm 500 million years ago (according to 3 out of 4 proxy measurements) to 280 ppm before the industrial revolution and 380 ppm today, i.e. CO_2 concentrations were much, much higher at the beginning of the Phanerozoic (542 Myr BP). More importantly, the beginning of the Phanerozoic coincides with an unprecedented explosion in the fossil record (the Cambrian Explosion) and this has serious implications for the modern vilification of CO_2. I must admit, it's not clear whether the explosion corresponds with an explosion in the number of potential fossils, or just an explosion in the ability to fossilise, but, either way, Life was definitely thriving and high CO_2 was definitely *not* a problem (see Figure 4).

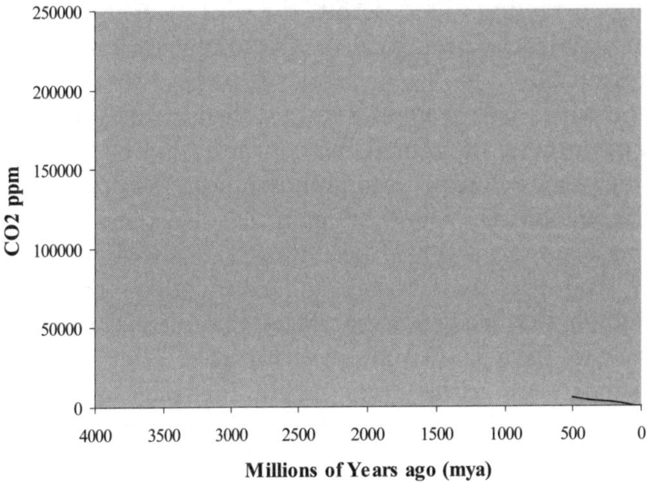

Figure 4 – *CO₂ over the last 500 million years (bottom right).*

It's true that the reliability of these educated guesses diminishes with time, but there are some universally accepted theories that do help us reconstruct even the very earliest embryonic climate (during the Hadean eon (3.8 – 4.6 billion years BP)). For instance, the Faint Young Sun Paradox has made the verifiable presence of liquid water on the young Earth hard to explain. According to the Standard Solar Model the Sun would have been approximately 30% weaker during Earth's primordial history, and, thus, would not have been strong enough to prevent water from freezing. But water *was* prevented from freezing, and one possible explanation is an extreme greenhouse climate (that traps a lot more of the Sun's heat) because of very high CO_2 levels.

This theory is supported by the logical source of Earth's early atmosphere, because it must have come from somewhere, and the most likely 'somewhere' is deep inside the planet itself. The chances are it was 'outgassed' by volcanoes and if Hadean volcanoes 'outgassed' like modern volcanoes, then primordial CO_2 concentrations would have been high enough to support liquid water, despite the faint young Sun, and an embryopaleoclimatological (?) CO2 range of 25-80% isn't unreasonable.

So what? This is about now, not then, right? What about photosynthesis though? If the concentration of CO_2 keeps falling, what will happen to photosynthesis? Basically, if CO_2 concentrations fall below 200 ppm most trees and broad leaf plants begin to struggle[*]. And if they fall below about 90 ppm they stop photosynthesising altogether; which means that dwindling CO_2 is a genuine problem. Commercial greenhouse managers have known about it for ages. They have been enriching plant growing atmospheres for years, to 1000 ppm and above, as Advance Greenhouses will now explain:

---

[*] Most trees and broad leaf plants use the C_3 photosynthesis pathway. Unfortunately this requires passive diffusion of CO_2 and can't cope with low concentrations.

'Carbon Dioxide (CO2) is one of the easiest ways to accelerate plant growth. Plants grown with supplemental CO2 can produce up to 40% more flowers or fruit. A propane or natural gas CO_2 generator is the most cost effective way to add CO_2 to your environment. Many greenhouses use CO_2 generators to boost CO_2 levels safely and economically.' [295]

Scott Denning, Ph.D., a physicist from Colorado State University in Fort Collins, Colorado and active member of the North American Carbon Program, remains typically confused:

"Stuff is growing faster than it's dying [because of increased CO_2 levels], which is weird.' [296]

But that's typical of people who can't see the plant fuel for the carbon dioxide.

The point is the current carbon dioxide concentration (380 ppm) doesn't support profitable greenhouses because it's limiting photosynthesis, as the IPCC well knows:

'Primary production[*] is carbon limited in terrestrial ecosystems in part because of (geologically speaking) low CO_2 concentrations.' [297]

Some plants have even evolved a way of concentrating carbon dioxide within their leaves to try and cope:

'Low CO2 may have been the stimulus that favoured the evolution of C_4 plants, which increased greatly in abundance between 7 and 5 Myr BP.' [298]

Basically, carbon dioxide concentrations are hovering just

---

* Primary production is, according to the IPCC: 'The amount that is 'fixed' from the atmosphere, i.e., converted from CO_2 to carbohydrate during photosynthesis'.

above a critical limit for non C_4 plants, not just below a critical level for the destruction of Earth.

Where has it all gone though? Well, it's funny you should ask that, because a lot of it was buried as decomposing organic carbon products. The truth is that dead plants (and animals) that are buried before they release their carbon end up as fossilised carbon stores, like oil and coal. In the words of the WWF:

> 'Four billion years ago its concentration in the atmosphere was much higher than today (80% compared to today's 0.03%), but most of it was removed through photosynthesis over time. All this carbon dioxide became locked in organisms and then minerals such as oil, coal and petroleum inside the Earth's crust.' [299]

Which means Life is responsible for removing the carbon dioxide the 'stewards' are now suggesting will kill...Life.

Brilliant.

Let's quickly summarise the story so far then: the concentration of carbon dioxide is limiting photosynthesis because it has dropped by 99.848 % but the 'stewards' want it to go even lower because they think it will kill a planet that has only ever known it higher.

But CO_2 is making the planet hotter, right? Well, it can't get much colder, because it's only just above 'ice age' at the moment. In fact, we're still in an ice age. Now is just a slightly warmer period (interglacial) between two slightly colder periods (glacials). In the words of the British Natural History Museum:

> '[Humans] are currently in an icehouse phase but for most of the last 600 million years the Earth has been much warmer than at present.' [300]

And the IPCC:

'There are strong indications that a warmer climate, with greatly reduced global ice cover and higher sea level, prevailed until around 3 million years ago. Hence, current warmth appears unusual in the context of the past millennia [particularly to Al Gore], but not unusual on longer time scales.' [301]

There is a caveat of course, because, as always, change can change things but only while the rate of change doesn't change:

'If projections of approximately 5°C warming in this century (the upper end of the range) are realised …there is no evidence that this rate of possible future global change was matched by any comparable global temperature increase of the last 50 million years.' [302]

But the last 50 million years does, rather conveniently, exclude a global temperature increase that *is* comparable, as the IPCC will now explain:

'Approximately 55 Ma [million years ago], an abrupt warming (in this case of the order of 1 to 10 kyr [thousand years]) by several degrees Celsius is indicated …The warming and associated environmental impact was felt at all latitudes, and in both the surface and deep ocean…The climate anomaly…occurred at the boundary between the Palaeocene and Eocene epochs, and is therefore often referred to as the Palaeocene-Eocene Thermal Maximum (PETM).' [303]

Did they deliberately exclude this from the previous statement to make things seem unprecedented? I don't know, but things *aren't* unprecedented, as they well know:

'The PETM was one of the most rapid and extreme global warming events recorded in geologic history.' [304]

Here are the thoughts of the National Geographic Magazine:

> 'Earth experienced its warmest spell since the age of the dinosaurs. Crocodiles basked in Greenland then, and Mediterranean warmth apparently reached the very top of the planet…Scientists think a powerful greenhouse effect from atmospheric carbon dioxide ten times higher than today's fuelled the warm spell.' [305]

And if that doesn't make the 'cataclysm invokers' think, nothing will.

Five million years later there was another climate event, this time in the opposite direction and this time caused by lots of little free-floating *Azolla* ferns. They blossomed in the warm (ish) Arctic Ocean so that 'fifty million years ago the now ice-locked northern ocean may have resembled a freshwater pond, choked with waterweed and teeming with microscopic life'[306]. They also trapped vast amounts of carbon dioxide at the bottom of the Arctic Ocean however and that may have triggered the current 'icehouse' phase, as conclusively proved by the Darwin Centre for Biogeology:

> 'Interestingly, around this same period a climatic transition occurred from greenhouse to icehouse.' [307]

OK, so I may have taken a few liberties with the 'conclusive proof', but you should hear what a team of scientists from the Universities of Cardiff, Bristol and Texas had to say:

> 'The link between declining CO_2 levels in the earth's atmosphere and the formation of the Antarctic ice caps some 34 million years ago has been confirmed for the first time in a major research study…The study's findings, published in Nature online, confirm that atmospheric CO2 declined during the Eocene - Oligocene climate transition and that the Antarctic ice sheet began

to form when CO_2 in the atmosphere reached a tipping point of around 760 parts per million (by volume).' [308]

i.e., low carbon dioxide levels and temperatures are linked and recent, 'climate on Earth has changed on all time scales, including long before human activity could have played a role'[309] and there's no reason why carbon dioxide levels, temperature and climate shouldn't change again.

Except if you really like familiar ecosystems of course.

MORE FASHION, AGAIN

What we are talking about, yet again, is fashion, because no matter how poetic the 'stewards' become there's still no objective reason to believe that global warming will kill the entire planet. Indeed, and as Professor Lovelock has previously pointed out, what the 'stewards' mean by the plea 'save the planet' is 'save the planet as we know it' and that means they're being prejudiced again. Why else is an 'icehouse' better than a 'hothouse' for instance, especially when you remember that 'Pre-Quaternary climates prior to 2.6 Ma [million years ago] were mostly warmer than today'[310]?

Perhaps Professor Lovelock can explain:

'Gaia seems to like it cold, which is why perhaps for most of the last two million years, and maybe much longer, the Earth has been in an ice age.' [311]

Professor Lovelock please. Two million years represents just 0.057% of Gaia's life, and just 0.37 % of the last 542 million years (the Phanerozoic eon).

He continues:

'I think it important to recognise that a hot Earth is a weakened Earth.' [312]

I assume that conclusion is based on the last two million years as well, but I would love to see him discuss 'weakness' with previous 'weak Earth' species, like…the Dinosaurs[*]. Make no mistake, Professor Lovelock does have all sorts of legitimate things to say about sterile oceans, spreading deserts and human consequences, but his basic implication remains the same:

> I think it important to recognise that dinosaurs were pansies.

He even states (in the next book admittedly) that 'the best known hothouse…marked the dawn ('eos') of large mammals'[313], i.e. hot weak Earth's can also be the dawn of whole new evolutionary dynasties. Remember, according to the IPCC, 'indicators for the presence of continental ice on Earth show that the planet was mostly ice-free during geologic history'[314], so claiming that 'a hot Earth is a weakened Earth' isn't based on observing nature, it's based on the fact that Professor Lovelock really liked 'the lush, comfortable and beautiful Earth we left behind some time in the twentieth century'[315]. It's subjective and all such habitat love is the same. It's not about the intrinsic value of life, it's about the extrinsic value of familiarity instead and that's *not* objective natural history. For instance, even though Azolla ferns really liked the Arctic Ocean when it wasn't frozen, the WWF really likes it when it is:

> 'The Arctic is no longer the pristine environment of our childhoods. While oil exploration contaminates the habitat, climate change has caused a local temperature rise of 4°C over the last 50 years. This may not sound like much, but it's been enough to melt an extra 1.3 million km² of ice between 1978 and 2005.'[316]

---

[*] Dinosaurs lived on Antarctica. That's how warm it was.

The first point I would like to make is that the Arctic is no longer the unfrozen pristine environment of the Palaeocene-Eocene Thermal Maximum either, but that's beside the point isn't it, because the WWF likes ice and that's that.

This isn't just about the icy 'pristine environment of our childhoods' though is it?

Of course it's not:

'With less ice, a polar bear finds it much harder to hunt. With no ice, a polar bear is unable to hunt at all.' [317]

To be fair, the WWF has probably managed to project its subjective extrinsic values onto the intrinsically neutral Arctic ice as well, but, more often than not, ecosystem love is about species love and this time it's 'the icon of the north'[318]:

'Unless these problems are tackled at once, the Arctic's summer ice may be lost by 2040, and the polar bear's extinction is predicted by the end of the century.' [319]

What they fail to point out is that most, if not all, the polar bears that make up 'the polar bear' will be dead by 2040 anyway, and that's based on their own lifespan predictions of '20 to 30 years'[320]. For those who haven't noticed, '20 to 30 years' means the youngest polar bear alive today will die in 2039 anyway, at the latest. That's not the point though is it, as long as the last one doesn't die, nothing else matters.

Perhaps the worst example of habitat racism involves the aquatic 'dead zones' though, because they aren't even dead. Before we explore the reasons why living life is now dead however, here's Greenpeace to explain why dead zones are considered dead at all:

'This situation is the result of coastal waters algae blooms due to nutrient pollution from sewage discharges and agriculture. As the algae die and rot, they consume oxygen, and the oxygen dissolved in the water falls to levels unable to sustain marine life.' [321]

Crumbs, 'unable to sustain marine life', that's a pretty definitive statement. It's supported by NASA though (although not strictly relevant to space exploration I must admit):

'Enhanced phytoplankton blooms can create dead zones. Dead zones are areas of water so devoid of oxygen that sea life cannot live there.' [322]

Blimey. I wonder what the National Science Foundation (NSF) thinks:

'Most sea creatures flee or suffocate to death in the Dead Zone's oxygen-starved waters, leaving highly adaptable jellyfish to proliferate unrestrained by predators and to gorge on the bounty of plankton.' [323]

Hang on. 'Jellyfish'? 'Gorging on a bounty of plankton'? That doesn't sound particularly 'dead' to me. Perhaps the University of Copenhagen can explain:

'Dead zones are low-oxygen areas in the ocean where higher life forms such as fish, crabs and clams are not able to live.' [324]

I see. So 'dead zones' aren't actually dead at all, they're just not full of 'higher' life forms. And what the 'stewards' actually mean by 'dead zones' is 'anaerobic primitive life zones' instead, which is undeniable habitat racism based on extrinsic attachments to 'fish, crabs and clams' rather than primitive jellyfish, and plankton.

The 'stewards' really don't like jellyfish do they.

Incidentally, some clams *can* live in dead zones after all, as discovered by Brown University:

'Coastal dead zones, an increasing concern to ecologists, the fishing industry and the public, may not be devoid of life after all. A Brown scientist has found that dead zones do indeed support marine life, and that at least one commercially valuable clam actually benefits from oxygen-depleted waters…Andrew Altieri…found that quahog clams (Mercenaria mercenaria) increased in number in hypoxic zones…The reasons appear to be twofold: The quahogs' natural ability to withstand oxygen-starved waters, coupled with their predators inability to survive in dead zones. The result: The quahog can not only survive, but in the absence of predators, can actually thrive.' [325]

Subjective, species-focused habitat love isn't just limited to cold areas and 'dead' zones though. It's rife throughout the conservation world and has even generated the active belief that humans are better at being nature than nature. The RSPB manages lot of reedbeds for example, even though it requires, in their own words:

'…preventing succession.' [326]

Preventing succession? Can you believe that? The RSPB champions itself as 'Nature's voice' while specifically and willfully 'preventing succession'. Honestly, what the hell is wrong with these people?

Here's the British Dragonfly Society (BDS) to explain why managing familiar habitats is so important:

'If left unmanaged, development of the vegetation and the encroachment of invasive plants will lead to the loss of [Scarce Blue-tailed Damselfly] colonies.' [327]

You guessed it, change threatens a species they really like. This time it's the Scarce Blue-tailed Damselfly:

'The aquatic vegetation needs to be managed to maintain early successional stages as the larvae live in substrates that support early successional plants. Therefore vegetation should be removed on a regular basis and aggressive species such as Greater Reedmace, Typha latifolia and rushes restricted.' [328]

Not the biggest fans of aggressive species such as Greater Reedmace, Typha latifolia and rushes are they.

The Herpetological Conservation Trust (HCT) is similarly prejudiced against woodland species:

'Heathland must remain open and warm in order to maintain its wildlife interest and therefore needs constant management to prevent succession to woodland.' [329]

And here's how they do it:

'Our management prescription includes the removal and subsequent control of tree cover…Today our efficiency is greatly improved through the use of specialised machinery wherever and whenever appropriate.' [330]

Crumbs, they even have specialised machinery to help them 'manage' (/fight) nature.

They don't like pond succession habitats much either:

'Ponds do not last forever, eventually they become dry land after being over grown by plants in the process of plant succession.' [331]

Perhaps they will accept nature's progress this time though:

'New ponds can be dug to replace ones that have been lost or old ones can be restored by removing the vegetation which has overwhelmed them…Many individuals

using hand tools, or fewer using power tools and machinery, can soon achieve the conservation objectives.' [332]

Perhaps not.
I must just repeat the last line quickly:

'Many individuals using hand tools, or fewer using
power tools and machinery, can soon achieve the conservation objectives.'

Remember, conservation objectives specifically contradict
nature's objectives and the HCT has thus perfectly illustrated
the difference between managing nature (conservation) and
observing nature (natural history).
You can even go on nature-fighting conservation holidays
if you want. Here's responsibletravel.com ('the world's leading
travel agent for responsible holidays') to explain:

'Valuable wildlife habitats are largely sustained through
human activity. Centuries of controlled grazing, mowing
or burning have produced specialist communities of
wildflowers, birds and invertebrates. The clearance of
encroaching vegetation helps to keep these precarious
communities in an early stage of succession preventing
the establishment of more competitive species. On our
holidays, you can get involved in a variety of different
activities from scrub clearance through to haymaking,
or removing poisonous ragwort.' [333]

It's totally anthropocentric habitat love. Perhaps Paul Evans,
an environmental correspondent for the Guardian, has said it
best though:

'Although it is not described in these terms, the battle
now is to protect the nature we like from the nature
we don't like.' [334]

Perfect.

It's critical to realise that none of this has anything to do with objective natural history though. This is blatant self-reverential habitat racism and it's all based on the assumption that nature is too pathetic to cope on its own. Nature is *not*, under any circumstances, too pathetic to cope on its own though. It has survived for 3.5 *billion* years because it's a ruthless warrior, not a baby mouse, and human attempts to help are anthropomorphic and anthropo-moronic; it's like offering a snorkel to a fish, or a ladder to a fruit bat, or a wet suit to a whale, or cutlery to a shark. In reality there's no doubt the planet will be different, but it will also be just fine.

Homo Sapiens on the other hand…

Chapter 9

UNSUSTAINABLE

By now you should really be wondering what conservation actually is, but we will come to that in the next chapter. Right now there's one last 'green' myth to kidnap and drop off a cliff and this time it's the holy grail of nature management. This time it's 'sustainability' because, apparently, Charles Darwin was wrong, and 'sustainability' is the new survival ability.

First point to note is that it's utterly impossible to sustainably harvest individuals. In fact, one of the major reasons the concept exists at all is because individuals have been subjectively grouped according to the extrinsic emotional impact of how they look, or what colour they are, or what they taste like, or what they like the taste of etc. It primarily exists because the 'stewards' think 'all = one', but it doesn't. In reality, 'all = all' and thus it should be face-punchingly obvious that individuals *can't* be sustainably harvested. For the benefit of anybody who skim reads, or subconsciously unreads things they don't like:

It is impossible to sustainably harvest individuals.

This debate extends beyond the intrinsic value of harvested individuals however, because we're now talking about the intrinsic sustainability of the *harvesters* themselves and that's a very different issue.

HARMONY

I should also point out that the unit of natural selection doesn't really matter anymore. It can be the gene or the individual, or your children, or your children's children, or even the species as a unit through space and time if you want, it really doesn't matter. The only thing it can't be is 'the planet' of course, or 'the ecosystem', because these are summaries, not selection units. At every other level there is competition to survive and that means elimination and population restrictions. Which is where the fallacy begins, because the 'stewards' have mistaken the struggle for life for intrinsic 'sustainability' and decided that nature is about calm meditation and 'harmony', rather than ruthless competition and violence. Here's the WWF to explain how perfectly lovely everything is for example:

> 'An ecosystem is balanced when the natural animals and plants and non-living components are in harmony- i.e. there is nothing to disturb the balance.' [335]

How tranquil. I wonder if Charles Darwin agreed:

> 'Battle within battle must be continually recurring...'

Clearly not, and I'm pretty sure a lot of animals wouldn't either. For instance, when a Harp Seal is caught by a Polar Bear does anybody really think it will be feeling 'harmonious'? Does anybody really think its 'balance' will remain undisturbed? And what about a Sambar Deer as it's being dragged to the ground by a tiger? Or a herring as it disappears down a whale's throat? Or a young breeding bird as it's being attacked by a magpie? Or a gazelle as it's being torn apart by African Wild Dogs? Or any other victims of nature's countless battles within battles? Does anybody really think that any of them will be feeling 'harmonious'?

Makes you wonder where the World Wide Fund *for* Nature is getting its World Wide Facts *about* Nature doesn't it. In reality, nature's a war, not a Sunday picnic, and most of the time 'harmony' will be the last thing on anything's mind. I have to admit, I have no idea what will actually be on their minds, but I can guarantee one thing: it definitely won't always be 'harmony'.

Harmonious aspirations are very definitely on the minds of the 'stewards' though, especially when it comes to *Homo sapiens,* and especially when it comes to Professor Lovelock:

> 'It may be that the destiny of mankind is to become tamed, so that the fierce, destructive, and greedy forces of tribalism and nationalism are fused into a compulsive urge to belong to the commonwealth of all creatures which constitutes Gaia.' [336]

Commonwealth of all creatures? What about the struggle for life? Nature isn't a democratic 'commonwealth', it's a winner-takes-all death match, where winners eventually lose all anyway. It's a war of survival and it's specifically based on 'fierce, destructive, and greedy' creatures because 'gentle, constructive, and selfless' creatures are always 'un-gently, caught and eaten'.

Carl Gustaf Lundin, the Head of the IUCN's Marine Programme, disagrees of course:

> 'Beneath the surface of the oceans lies an extraordinarily diverse and relatively peaceful world.' [337]

But I would love to watch him explain how 'peaceful' everything is to a herring baitball as it's being destroyed by ocean predators, or a seal pup as it's being hit by a hungry Great White Shark, or a squid as it's caught by a Sperm Whale.

Peaceful indeed.

The WWF has even made human 'harmony' its central aim:

> 'The mission of WWF is to stop the degradation of the planet's natural environment and to build a future in which humans live in harmony with nature.' [338]

As has Flora and Fauna International:

'We are driven by a vision of a healthy planet where people and wildlife can co-exist harmoniously.' [339]

And the Wildlife Conservation Society:

'Wildlife Conservation Society…activities change attitudes towards nature and help people imagine wildlife and people living in harmony.' [340]

And Friends of The Earth:

'Our vision is of a peaceful and sustainable world based on societies living in *harmony* with nature.' [341]

And Conservation International:

'Our mission is to conserve the Earth's living heritage – our global biodiversity – and to demonstrate that human societies are able to live harmoniously with nature.' [342]

And the International Union for the Conservation of Nature:

'Our vision and mission remain relevant in a rapidly changing world, but if we are to meet these ambitious aims we need to vastly expand IUCN's ability to influence change to enable humankind to live sustainably, in harmony with the natural world.' [343]

And the RSPCA (although I have absolutely no idea how they can say this and endorse commercial broiler chicken production at the same time):

'The RSPCA's vision is to work for a world in which all humans respect and live in harmony with all other members of the animal kingdom.' [344]

And the British Natural History Museum:

> 'Conservation projects across the globe are learning more about how our world works and how we can live in better harmony with it.' [345]

In fact, the concept defines modern conservation even though 'harmony' is a subjective utopian desire that has absolutely nothing to do with objective natural history. In the words of Charles Darwin:

> 'All that we can do, is to keep steadily in mind that each organic being is striving to increase in a geometrical ratio; that each at some period of its life, during some season of the year, during each generation or at intervals, has to struggle for life and to suffer great destruction.'

I will assume that everybody understands what he meant by 'great destruction', and that everybody is also happy to accept that harmony and 'great destruction' aren't the same thing, but for those who aren't sure what Charles Darwin meant by a 'geometrical ratio', here's…Charles Darwin to explain:

> 'The elephant is reckoned the slowest breeder of all known animals, and I have taken some pains to estimate its probable minimum rate of natural increase; it will be safest to assume that it begins breeding when thirty years old, and goes on breeding till ninety years old, bringing forth six young in the interval, and surviving till one hundred years old; if this be so, after a period of from 740 to 750 years there would be nearly nineteen million elephants alive, descended from the first pair.'

For those who are still unsure, a geometric ratio is a sequence of numbers where each subsequent number is a consistent multiplication of the previous number. For example, based on doubling the previous number each time, 2 becomes 4 which becomes 8, then 16, 32, 64, 128, 256, 512, 1024, 2048 and so on. In the world of nature it refers to the number of descendants that could accumulate if all offspring survived, i.e. two elephants could become nineteen million elephants in 740 to 750 years.

The 'stewards' will field one apparent exception to this rule of course, because affluent humans don't produce enough children to maintain the population, never mind increase it in a 'geometrical ratio', but that doesn't mean they're not *trying* to expand. It just means consumption is growing faster than reproductive capacity and that contraception is causing 'great destruction' before any children are born. Indeed, affluent humans haven't stopped having sex or even conceiving in many cases. They've just decided the period of their life where they suffer 'great destruction' is around conception rather than after birth. It's a little bit like a high infant mortality rate, but it doesn't mean the population isn't trying to expand. It just means the population *can't* expand, because of relative resource restrictions.

Which brings me to the subject of resources, because resource competition defines populations, not benevolence and love. Remember, 'each organic being is striving to increase in a geometrical ratio', and that means each organic being is also striving to increase resource consumption in a geometrical ratio. For instance, in predator-prey relationships more prey doesn't mean predators start making biltong and jerky; it means more food, better survival rates and ultimately, more predators. Conversely, when over-hunting or over-grazing forces prey numbers down again, predator numbers must follow. The two rise and fall with their respective resources (see Figure 5) in an apparently harmonious mathematical relationship that's actually based on limited resources, not limited appetite.

Please note, each downward slope of the graph means the

death rate has exceeded the birth rate. And that means 'harmony' is probably not a description that starving animals would necessarily agree with.

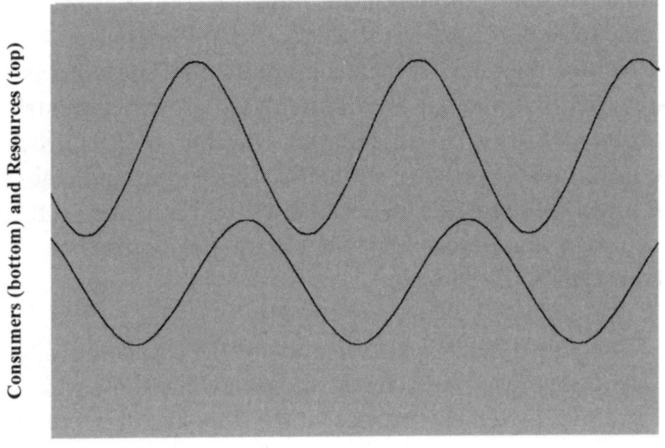

Figure 5 – *The 'harmonious' increasing and decreasing of consumers and their resources.*

Such perfect harmony between consumers and resources rarely if ever happens in the wild of course, even in isolated predator-prey relationships such as the wolves and moose of Isla Royale, but populations *are* logically limited by resources and the conclusion remains.

It even applies to biodiversity in general. When conditions are favourable for instance, biodiversity expands. And when conditions are unfavourable, biodiversity contracts. It isn't limited by wise prudence and concerns about sustainability though, it's limited by exhausted resources and the conditions of life instead. And if a massive asteroid hits the planet, or a supervolcano erupts, or a fossil-fuel-based super-exploiter evolves, it's severely limited.

Life keeps on trying to expand though. It may abandon fragile, highly-sensitive species, like Dodos, Baijis (Yangtze River Dolphins)[*] and Giant Pandas, in favour of robust, highly-resilient species, like Grey Squirrels, jellyfish and Cane Toads, but that's a function of limited resources, not limited intent. In fact, regardless of whether there's resource growth or resource recession, if there's more than *one* descendent per parent per lifetime that means attempted growth and intrinsic *unsustainability*. It doesn't matter whether species produce thousands of offspring (r-strategists, like turtles and crabs) or just a few (K-strategists, like wildebeest and humans), if there's more than *one* descendent per parent per lifetime they are intrinsically *unsustainable*:

> 'The only difference between organisms which annually produce eggs or seeds by the thousand, and those which produce extremely few, is, that the slow-breeders would require a few more years to people, under favourable conditions, a whole district, let it be ever so large.'[Darwin]

Imagine how many dandelions there would be if every seed of every dandelion consistently survived to reproduce for example. And imagine how many frogs there would be if every tadpole of every frog consistently survived to reproduce. And what about Comb Jellyfish? Apart from being a second rate invasive 'pest' in the eyes of the 'stewards', it took them just 10 years to account for 90% of the biomass of the Black Sea and that's clearly unsustainable (and incredible) population growth. And what about Cane Toads in Australia, or Grey Squirrels in Britain? These aren't just examples of Life's expedient brilliance, they're also examples of it's staggering geometric potential as well.

---

* The Baiji was declared functionally extinct in 2006 after a survey of the Yangtze river in China failed to find a single specimen. There has been an unconfirmed sighting since that survey, but isolated individuals won't be able to save the species even if they do exist.

Here's David Klein, of the University of Alaska, to explain how sustainable twenty nine Reindeer can be under the right conditions:

'Reindeer (Rangifer tarandus) introduced to St. Matthew Island in 1944, increased from 29 animals at that time to 6,000 in the summer of 1963…Increasing at geometric rates, the populations passed from moderate levels, with respect to the food supply, to excessive populations in only a few years.' [346]

Which led to:

'Overgrazing of lichens on the island…'

And thus a:

'…crash die-off the following winter to less than 50 animals.'

Due to:

'…starvation…'

In other words reindeer aren't intrinsically sustainable either.

Even tigers are the same. According to the WWF, female 'tigers generally attain sexual maturity at 3-4 years…give birth to 2 to 3 cubs every 2 to 2.5 years…[and] have been known to reach the age of 26 in the wild'[347], which clearly means a lot more than one descendent per parent per lifetime and, if there were no limits on offspring survival, they would quickly demonstrate how intrinsically unsustainable they are.

'Lighten any check, mitigate the destruction ever so little, and the number of the species will almost instantaneously increase to any amount.'[Darwin]

And every other species I have previously mentioned in this book is *exactly* the same. As is every other species that has ever existed. They're all trying to expand as much as possible and they're *all*, without exception, and at the risk of over repeating myself, intrinsically *unsustainable*. And rightly so, because Life *can't* fight to survive if organic beings *don't* fight to succeed. In reality, unsustainability is the fuel of evolution itself. It maintains the competition needed to keep the system vigorous and healthy and it remains a vital part of nature therefore, regardless of what the 'stewards' think. In fact, the 'harmonious' sustainability that the 'stewards' see isn't 'harmonious' sustainability at all, it's precariously balanced violent *unsustainability* instead, and that's a *massive* misconception that must surely sever all remaining links with objective natural history:

> 'In looking at nature, it is most necessary to keep the foregoing considerations always in mind—never to forget that every single organic being may be said to be striving to the utmost to increase in numbers; that each lives by a struggle at some period of its life; that heavy destruction inevitably falls either on the young or old, during each generation or at recurrent intervals.'
> Darwin

One more time, for the benefit of everybody who still thinks that nature is about love and friendship: '…every single organic being may be said to be striving to the utmost to increase in numbers'[Darwin] and, as a result, they're all intrinsically *unsustainable*.

And that includes humans.

BOOM

Consider the graph in Figure 6 (overeleaf). It shows the human population over the last 2,000 years and it should speak for itself:

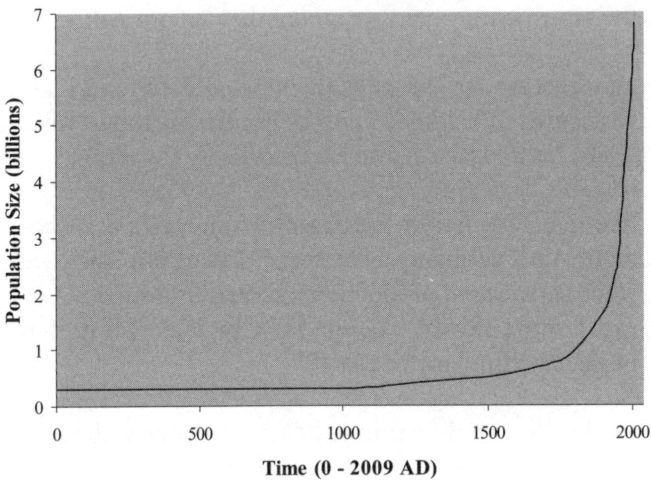

Figure 6 – *The human population over the last 2000 years*[348].

Does anybody really think humans can live in 'harmony' with nature?

According to the United Nations Population Division, the population hasn't even stopped growing yet:

> 'World population is projected to reach 7 billion early in 2012, up from the current 6.8 billion, and surpass 9 billion people by 2050.' [349]

That's 6.8 billion people now, 7 billion by 2012 and a staggering 9 Billion by 2050. I would suggest that if the 'stewards' think there are problems now, they will be drinking hot acid by 2050. They steadfastly refuse to stop supporting continued growth though. For example, the National Geographic Magazine openly recognises the problem:

> 'We don't want to think about limits. But as we approach nine billion people on the planet, all clamouring for the same opportunities, the same lifestyles,

the same hamburgers, we ignore them at our risk.' [350]

Happily agrees with 'perhaps the most vilified [and realistic] social scientist in history', Thomas Robert Malthus[*] (who also inspired Charles Darwin and his theories by the way):

> 'Human population, he observed, increases at a geometric rate, doubling about every 25 years if unchecked, while agricultural production increases arithmetically—much more slowly. Therein lay a biological trap that humanity could never escape.' [351]

Willingly accepts that things haven't changed, despite the agricultural advances of the twentieth century ('the green revolution'):

> '…hunger, famine, and disease are with us still, just as Malthus said they would be.' [352]

Indirectly accepts the nipple-pinchingly obvious solution:

> 'Every time food supplies plateaued, population eventually levelled off.' [353]

Specifically quotes a realistic professor of population studies (Tim Dyson from the London School of Economics):

> 'People who say Malthus is wrong usually haven't read him…No one in their right mind doubts the idea that populations have to live within their resource base. And that the capacity of society to increase resources

---

* Thomas Malthus wrote *An Essay on the Principle of Population* (1798), which included the patently ridiculous idea that populations can grow faster than resources and thus will always be pressed against their resources limits and thus will always have poverty of some sort.

from that base is ultimately limited…And this notion that we can grow forever, well it's ridiculous.' [354]

And then, quite remarkably, rejects the whole lot in favour of supporting further population growth:

'To meet rising food demand, we need another green revolution, and we need it in half the time.' [355]

Brilliant.

National Geographic Magazine described this as a 'Special Report' by the way, but that's clearly a report with a capital 'SPECIAL'. Essentially, they have recognised that population growth threatens survival via inadequate resources, and that increased resources just supports further growth that eventually re-threatens survival in exactly the same way, and then decided that continued population growth based on increased resources is the most sensible option anyway.

Here are the thoughts of Richard Dawkins:

'It is no use appealing to advances in agricultural science – 'green revolutions' and the like. Increases in food production may temporarily alleviate the problem, but it is mathematically certain that they cannot be a long-term solution; indeed, like the medical advances that have precipitated the crisis, they may well make the problem worse, by speeding up the rate of the popula-tion expansion.' [356]

I have no idea how National Geographic Magazine plans to address the next 'global food crisis' then, but that's beside the point, because their refusal to accept that continued pop-ulation growth is the cause of the problems they're trying to address is actually an agreement to unconsciously accept that striving to increase in a geometrical ratio is the basis of all life on Earth. Which is ultimately the point I was trying to make in the first place.

Professor Lovelock thinks that striving to increase in a geometrical ratio is pure folly of course:

> 'Our inherited urge to be fruitful and multiply, and to ensure that our own tribe rules the Earth, thwarts our best intentions.' [357]

He does try to make us feel better as well:

> 'It is absurd to expect us to change ourselves as it would be to expect crocodiles or sharks to become, through some great act of will, vegetarian. We cannot alter our natures, and as we shall see the bred-in tribalism and nationalism we pretend to deplore is the amplifier that makes us powerful.' [358]

But, apart from condemning *himself* for pretending to deplore us, what Professor Lovelock has failed to realise, and what National Geographic Magazine has unconsciously accepted, is that being fruitful and multiplying is the whole point. Imagine how far Life would get being fruitless and un-multiplied.

Now, I'm not saying that will make the consequences any easier. I'm just saying that all species on Earth are trying to be fruitful and multiplied and that our attempt to do so does *not*, under any circumstances, make us different or wicked. The 'stewards' may hate us for being unsustainable planet rapists, but we're just trying to be as successful as possible with the time and resources available, just like every other species that has ever existed. Using modern medicine and agriculture etc., humans have done a pretty good job as well:

> 'In such cases, and endless others could be given, no one supposes that the fertility of the animals or plants has been suddenly and temporarily increased...The obvious explanation is that the conditions of life have

been highly favourable, and that there has consequently been less destruction of the old and young, and that nearly all the young have been enabled to breed.'[Darwin]

We could be doing a much *better* job of course, but the sun is (or has been) shining and, like Daphnia (water fleas) in an algal bloom, or frogs in the wet season, or farmers in the summer, humans have been making as much metaphorical hay as possible. As well as a lot of literal hay, and all the other wonderful things that *Homo sapiens* make. It's dramatic and unharmonious but that's the way of the world. Nature isn't a charity coffee morning, or a 'commonwealth', or a children's tea party; it's a struggle for life. It's a war and every organic being must take whatever opportunities it can.

Humans are upsetting the balance of nature though, right? Yes, they are, but nature's balance is always being upset, as previously discussed. In fact, if nature's balance actually got upset when it was upset, it would be spitting wasps by now. In reality, harmony is an illusion that has nothing to do with the struggle for life and a quick glance at a few newspapers will rapidly demonstrate the real human priority. Amongst endless stories about the environmental horrors associated with too many lives (overpopulation), you will find endless more about the glory of saving even *more* lives, even though the two are logically incompatible. For example:

'Humans are using 30% more resources than the Earth can replenish each year…The problem is also getting worse as populations and consumption keep growing faster than technology finds new ways of expanding what can be produced from the natural world…by 2030, if nothing changes, mankind would need two planets to sustain its lifestyle.' [359]

Can be found alongside:

> 'Heart disease is one of the world's leading killers, but now…we have hope of reducing the burden that the disease causes.' [360]

Either you want a lot of nature *or* you want a lot of humans though. You can't have both and deep down humans are the main priority, regardless of whether they're sustainable or not. First the people you know (which is why some people will spend £120,000 on one related premature baby instead of £120,000 on 10,300 unrelated unvaccinated babies[*]), then the people you don't know and it means that, just like every other species on Earth, utilisation is more important than preservation.

Population growth can't continue forever though, because infinite growth simply isn't possible. In fact, let me repeat that:

Infinite population growth is impossible.

Just like growth without impact isn't possible. In fact, perhaps I should repeat that as well:

Growth without impact is impossible.

Luckily, intrinsic unsustainability isn't just the quality that fuels Life's staggering vigour, it's also the safety net that prevents its self-destruction, because it means growing populations will collapse before they destroy the resources that support them, as the reindeer of St. Matthew Island clearly demonstrated. They reduced the lichen biomass so dramatically they eliminated themselves, but, critically, they didn't eliminate the lichen. Basically, because all organic beings are intrinsically unsustainable, they're also intrinsically self-limiting

---

* According to the United Nations Children's Fund (UNICEF), the basic cost of a lifetime vaccination package, which includes vaccinations against measles, mumps, rubella, polio, tetanus and meningitis, as well as the tools to administer the vaccines, totals just £11.65 ($17). That means the costs involved for one premature baby (£120,000) could vaccinate 10,300 children for life.

and that's another magnificent example of Life's unsentimental and totally ruthless expedient brilliance, even if it's not a particularly appealing prospect for humans, or the British Natural History Museum:

> 'We are continually using resources from the natural world to meet our short term needs. In doing so, we are in danger of destroying the ecosystems that support every living thing – including us.' [361]

Deep down even the 'stewards' must realise that infinite growth isn't possible though, and that Darwin's observation probably won't go away:

> 'There is no exception to the rule that every organic being naturally increases at so high a rate that, if not destroyed, the earth would soon be covered by the progeny of a single pair.'

Perhaps the humble petri-dish can offer an insight into the phases that have gone before and the phases yet to come (see Figure 7):

- ADAPTATION - During the adaptation, or lag phase, bacteria adapt to their condition of life. Population growth is minimal or absent and resources remain plentiful.
- EXPONENTIONAL GROWTH - During the exponential, or log phase, the bacterial population grows (birth > death). Population growth is rapid because they are well adapted and resources are plentiful.
- STASIS - During the stasis phase the bacterial population doesn't grow (birth = death). Growth stops because resources aren't plentiful anymore.
- DEATH – During the death phase bacteria die (birth < death). Population growth reverses because resources have been exhausted.

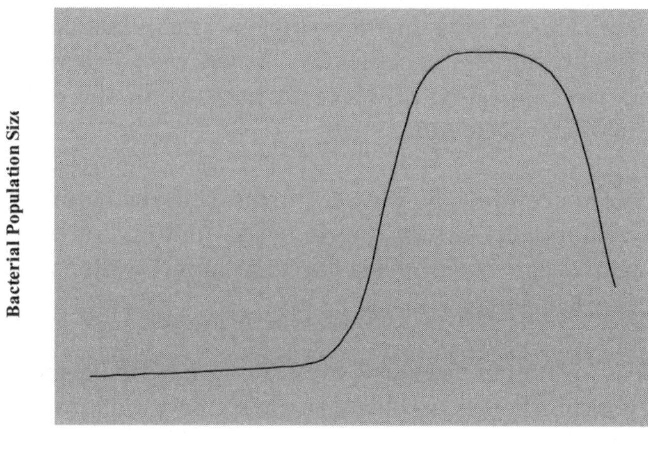

Figure 7 – *A typical bacterial growth curve.*

And maybe we should have a look at the world population graph again:

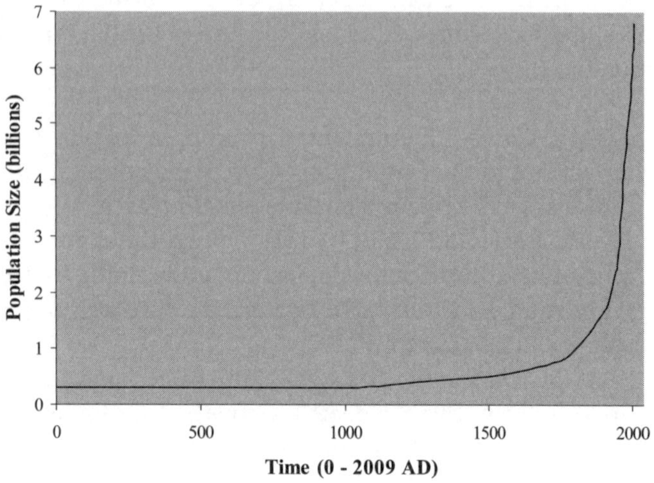

And then another look at the 'harmonious' consumer-resource graph:

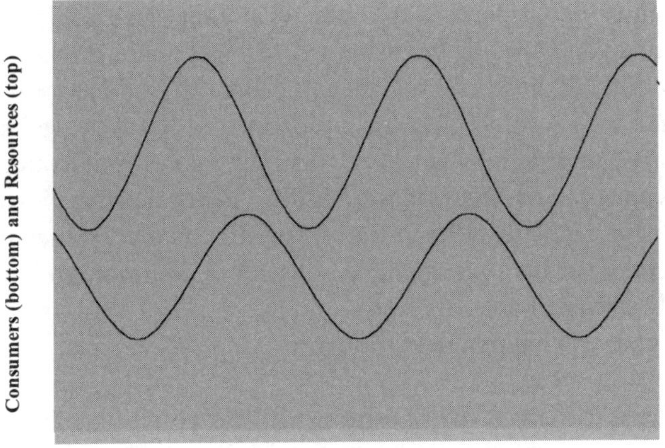

And quickly review current United Nation population forecasts:

'WORLD POPULATION TO EXCEED 9 BILLION BY 2050.'

I don't know about you, but it seems perfectly obvious to me that humanity has completed the adaptation, or lag phase, and is now charging madly up the slope of exponential growth, just like bacteria on a petri dish, or consumers in a resource feast. There's nothing wrong with that of course, because it's just the way life is. In fact, just like every other species that has ever lived on Planet Earth, humans are truly amazing because of it. Despite being utterly useless at being objective, they're miraculously fabulous at totally unsustainable super-survival and that's amazing.

They have done it all themselves too. Some may have a pathological desperation to give all the credit to remote creators who refuse to prove they exist, but in reality they alone have achieved everything they have ever achieved, same as all species. Nobody has done it for them. Nobody has cleared the way. They have done it all themselves, and when I think about what humanity has been through and what they have achieved I am, once and for all, *not* ashamed to be human, despite the best efforts of those who malign and belittle us (the 'stewards').

We may be changing the conditions of life, but we have been through hell to get here and that makes me proud. It takes all sorts: compassionate, ruthless, benevolent, malicious, idealistic, realistic, clever, stupid, theistic, secular, subjective, objective, yet here we stand, divided by a common goal and totally unsustainable, as always.

What will happen next though?

> 'Look closely at the natural world and you'll find that it is in constant flux. As the availability of energy and nutrients changes, and other environmental changes happen, populations rise and fall.' [362]

I will leave it up to you to decide.

Chapter 10

'THE HUMAN'

It should be perfectly obvious by now that conservation isn't natural history. If it isn't, here's a quick summary of the conclusions so far.

There are:

- no objective reasons to discriminate between ecosystems.
- no objective reasons to discriminate between species.
- no objective reasons to discriminate between individuals.
- no objective reasons to discriminate between dead animals.
- no objective reasons to worry about dead animals.
- no objective reasons to vilify unsustainability.
- no objective reasons for turning 'all' into 'one'.
- no objective reasons to worry about Life.

If you still think conservation is natural history after that, then I can only assume you are biologically preprogrammed to reject evidence in favour of prejudice, and I will accept defeat safe in the knowledge that I was never going to convince you anyway. For everybody else it should be quite clear that conservation just isn't natural history.

What on Earth is it then? What do you call something that uses subjective belief to promote guilt while simultaneously offering salvation through penance and faith?

MAGNIFICENTLY SIMILARLY DIFFERENT

Before answering that question, I must first make sure that no subjective racism remains. This is ground we have already covered I hear you say, but what about humans? Have you really accepted that each human is just another individual of just another species? Remember, this isn't about ethics or morality, this is just about the objective appraisal of humans in light of the evolutionary process they share with all other living organisms.

Many of you will vehemently disagree at this point, and you would not be alone. A survey of 2060 UK adults entitled *Faith and Darwin: Faith, Harmony, or Confusion*[*], found that just 14% of people believe that 'humans are just another species of animal and have no unique value or significance'[363], which means a staggering 86% believe humans are special.

This is only mildly skewed by religious proclivities as well (young earth creationists, intelligent design theorists etc.). When they focused on atheistic evolutionists in particular just 28% believe that 'humans are just another species of animal and have no unique value or significance', which means a staggeringly amazing 72% of atheistic evolutionists believe that humans are special. *72%*. But how? If there's no objective basis on which to elevate one species over another, why do 72% of atheistic evolutionists believe *our* species is better than every other? Counter-intuitive isn't it, but that's what's happening. Even Professor Lovelock has found a way of lifting us out of objective evolutionary obscurity:

> 'We are creatures of Darwinian evolution, a transient species with a limited lifespan, as were all our numerous distant ancestors. But, unlike almost everything before we emerged on the planet, we are also intelligent social animals with the possibility of evolving to become a wiser and more intelligent animal, one that might have a greater potential as a partner for the rest of life on Earth.'[364]

----

* *Faith and Darwin: Faith, Harmony, or Confusion* was commissioned by the theology think tank Theos, and conducted by the independent polling company ComRes.

Apart from the preposterous idea that Life on Earth needs a partner, it would appear that our 'transient species with a limited lifespan' is a supreme intelligence with a divine future and that:

'…we humans are vitally important as a part of Gaia.' [365]

Because:

'Humans could make the whole planet intelligent.' [366]

And therefore:

'Our obligation as an intelligent species is to survive.' [367]

I guess that means human intelligence is transcendent:

'Our intelligence is not something transcendent.' [368]

Oh, I guess not then. But isn't intelligence 'the ultimate weapon that lets us rule the world'[369]?

'It still allows us to be important, but the Earth can proceed without us, whereas without photosynthesisers it would probably soon die.' [370]

I see.

I think what Professor Lovelock is trying to say is that intelligence is vitally important in a less than vital, non-important type of way. Whatever he's trying to say though, the point is you can't objectively compare intelligence with totally different things, because better and worse, or important and more important, are subjective concepts based on relative perspectives. They may be appropriate when different species are judged against the same scale (i.e. which species can run the fastest), but they are definitely not appropriate when judging different species against different scales (i.e. intelligence vs. photosynthesis). For instance, humans can climb to great heights[*] and Sperm Whales[†] can swim to great depths, but which is better? Both species excel on different scales, but is climbing high objectively better than swimming deep?

No, it's not, and swimming deep isn't objectively better than climbing high either. Mountaineering just is and so is deep sea diving. They're different and comparisons between them are subjective. 'Bulldogs are better than healthy dogs' is another example. If it was declaring that Bulldogs are better because they're unhealthier than healthy dogs it would have an objective reference point and a legitimate claim, but it's not. Instead it's telling us about the subjective value of grotesque life-threatening deformity to that particular health-hating-paedomorphile. Similarly, Chris Packham's 'more valuable breeding adults' reveals more about his kneejerk ornithological prejudice that the intrinsic values of various different birds. 'Aboriginal subsistence whaling is better than commercial whaling' is another example, and: 'the Holocene extinction event [now] is worse than The Great Dying'. They're all based on personal prejudice and relative scale and the same applies to human self-reverence. Humans are definitely different, but how do 83% of humans in general, and a staggeringly amazing 72% of atheistic evolutionists, believe that humans are differently different?

Aware that at least 72% of you will be struggling right now, I'm sure someone will mention the soul, but nobody can

---

[*] By 'humans' I mean brave pioneers like Edmund Hillary and Tenzing Norgay rather than cowardly armchair pioneers, like me.

[†] I have to mention the ongoing debate about whether the Sperm Whale should be called *Physeter macrocephalus*, or *Physeter catodon.* Both were published by Linnaeus (the father of modern taxonomy) and, because the International Code for Zoological Nomenclature (ICZN) principle of 'First Reviser' means something about some stuff, that means some people disagree with some other people; and the sperm whale should really by classified as '*Physeter macrocephalus or Physeter catodon*'. Which seems sensible enough to me.

actually prove they have a soul. And nobody can prove that Pigeons, or Greenish Puffleg Hummingbirds, or Water Boatman, or Red River Hogs, or Giant Redwood trees, or any other individuals of any other species, don't. There's no evidence. You can know it's true in your heart, or your bladder, or your left nostril, or wherever the hell you like; it doesn't mean it's true. It just means that faith and common *nonsense* feels better than evidence and common *sense*.

The point is that difference is not actually a difference at all; it's a similarity. It's a consistent part of nature and anybody who can identify a non-different species is clearly looking through opaque-tinted spectacles. Biological taxonomists are trying really hard, but even they have had to find actual differences to separate different forms. They may have had to go sub-microscopic to find them, but they're there nonetheless and that's the point: non-different species aren't different, but only because they're the same. Humans are different, Spotted Wobbegongs[*] are different, Boomslangs[†] are different, Sea Angels[‡] are different, Vampire-Squids-From-Hell[§] are different. They're all different and they're all different because that's the whole point. Remember, even though the 'world's best natural history magazine' (BBC Wildlife) believes that species need to be protected from environmental change, that's *not* how the system works. Life probes its way through environmental change by constantly splintering in all directions and difference is a fundamental part of life. In the words of Professor Richard Dawkins:

'It is differences that matter in the competitive struggle to survive.'[371]

---

[*] The Spotted Wobbegong is a species of Carpet Shark found in the Indian Ocean.
[†] The Boomslang is a species of venomous snake.
[‡] Sea angels are small swimming sea slugs. They're also known as Cliones.
[§] In the sometimes bizarre world of binomial nomenclature, the Vampire Squid's latin name (*Vampyroteuthis infernalis*) literally means 'vampire squid from hell'. It's not actually from hell though; it's from temperate and tropical oceans instead.

They're essential and our continuing need to credit ourselves with the world's most different difference is subjective lunacy. Humans exist for the same reason as Fiddler Crabs, or Pacific Hagfish, or Aardvarks, or any other creature you care to think of. They exist because they fit the prevailing conditions. And that means humans aren't divinely differently different, they're magnificently similarly different instead.

GUILT + PENANCE + FAITH = SALVATION

Back to the question then:

> What do you call something that uses subjective belief to promote guilt while simultaneously offering salvation through penance and faith?

I should probably qualify the question before answering it though, and I will start with the use of subjective belief to promote guilt. I shouldn't have to spend too long on that however, because you probably already know how guilty you feel and, having read the proceeding chapters, you should hopefully know why you don't need to anymore. Life isn't defined by the last individual of each species, a warm Earth isn't a dead Earth, unsustainability isn't a biological crime, nature isn't about harmony, and so on. These are subjective beliefs, not objective observations, and they have nothing to do with the fundamental reality of life on Earth.

That's what we've been told though:

> 'In most case, the demise of a place or species can be directly attributable to one of humankind's many sins against the planet.' [372]

And that's why the 'stewards' continue to request financial penance for our 'sins' in return for promises that can't be tested. For instance, Flora and Fauna International (FFI), the world's first international conservation organisation states as fact that:

> 'The future of the Planet is in your hands, please don't throw it away.' [373]

But how do they know? How do they know that the future of an entire planet rests in your hands?

Basically, they don't. They've made it up. They've invented a doomsday scenario and then blamed it all on you so that you will feel guilty enough to seek redemption through financial penance. Which is where the faith comes in (apart from the faith needed to accept that the future of the planet rest is in your hands in the first place of course), because how do you know that the future of the planet will be safely 'in your hands' even if you do give money? You don't. That requires faith. Essentially, they have used subjective belief to promote guilt while simultaneously offering salvation through penance and faith.

They continue:

> 'The natural world is in crisis. Climate change. Rampant habitat destruction. Rapid population growth. Extreme poverty in many areas of the world. Global food shortages. One in four mammals threatened with extinction…The solution is in your hands.' [374]

I'm sure you spotted the guilt trip. And I have absolutely no doubt you can work out what the penance will be:

> 'To set up your monthly direct debit....'

And here comes the faith:

> 'We guarantee that your support will make a real difference.' [375]

How can they guarantee anything though? And what exactly are they guaranteeing? Surely they don't mean the end of climate change, and rampant habitat destruction, and rapid population growth, and extreme poverty in many areas of the world, and global food shortages, and extinction threats etc. etc.?

If they're promising salvation, I guess they do, and I'm sure they really believe it too. In the words of Richard Dawkins:

> '...these people actually believe what they say they believe.' [376]

OK, so Professor Dawkins was talking about religious fundamentalists, rather than wildly optimistic nature 'managers', but faith is faith regardless of what it's making up on the spot.

The WWF has offered a similar promise about the future of 'the polar bear':

> 'The Arctic's very existence is threatened by climate change…increasing CO_2 emissions and environmental pollution are triggering unprecedented damage to this once pristine habitat [guilt]…As a supporter of the Arctic Programme your donation will be used to help fund our Arctic and climate change work, which will in turn help protect the polar bear [penance]...We know our plans will work [faith].' [377]

How can they possibly know whether their plans will work though?

They can't. They're making it up. They're trying to convince you that an investment will help satisfy the guilt they have created, even though they have no idea whether it will actually work or not.

They have also promised to protect the future of the Orang-utan as well:

> 'The rainforests of Borneo are being lost at a staggering

speed…These rainforests are home to thousands of people and species. One of the most at risk is the orang-utan. Without our protection they could become extinct within the next 30 years [guilt]…Invest now to save the last home of the orang-utan [penance]…We know our plans will work [faith].' [378]

But do I really need to point out that 'the orang-utan' is an anthropomorphic term that totally ignores the fact that most individual Orang-utans will die over the next 30 years anyway? And do I really need to point out that 'the Orang-utan' they're referring to is actually 'the Bornean Orang-utan' (rather than 'the Sumatran Orang-utan') and that, as previously pointed out by media 'super-steward' Mark Carwardine:

'Species-splitting reduces population sizes and inflates the number of endangered species, painting an altogether bleaker picture.' [379]

In case you're wondering, and according to the 'world's foremost orangutan conservation organisation'[380] (the Orang-utan Foundation), the two were separated in 2004 because of 'their geographical distribution' and a few 'slightly different physical characteristics'[381]. These include facial narrowness, beard length, coat darkness, male cheek pad wideness and, quite splendidly, the amount of fruit they eat. To be fair to the WWF, 'the Sumatran Orang-utan' is even rarer, but that's not the point. However rare they may be the WWF is still generating guilt using a species that has been split by criteria that would shatter the human species into a thousand pieces, and it's still using that guilt to ask for money in return for salvation based on faith.

They even resort to insinuating blame if you don't donate:

'We can't do it without you…We need your help to make it happen.' [382]

i.e. if 'the Bornean Orung-utan' dies and you haven't donated it will be *your* fault. How does the WWF actually know they can save 'the Orang-utan' from extinction though?

The truth is, they don't. They think they can, but some people think Orang-utans were made in Heaven so that's hardly a guarantee. Indeed, it's not a guarantee:

> 'The anthropomorphus apes [i.e. Orang-utans] will no doubt be exterminated.'[Darwin*]

But they still 'know their plans will work' and they still make eternal promises.

Here's Greenpeace, on your involvement in their plan to solve everything:

> 'Join Greenpeace today and add your voice to the movement that's committed to defending our planet [against humanity - guilt]. Your support [penance] will make all the difference [faith].' [383]

But again, how do they know that 'your support will make all the difference'?

And what about the British Natural History Museum on 'fragile' ecosystems and species:

> 'Please donate now [penance] to help Museum scientists gain new knowledge to protect [faith] fragile ecosystems and sustain vulnerable species around the world [guilt].' [384]

And what about the Wildlife Conservation Society (WCS) on global biodiversity:

> 'Of all the species that share the earth, imagine one in

---

* Charles Darwin. *The Descent of Man* (1871).

five gone in the next decade. Some even estimate that 25,000 species might disappear in the next year alone. Hard to believe, isn't it? [guilt]…By supporting WCS [penance], you'll play a crucial role in the success of over 300 international field projects and important initiatives here at our Bronx Zoo headquarters [where you can "see tigers do puzzles" remember]…With your help, we will succeed in saving wildlife [faith].' [385]

I feel obliged to point out that almost all of the individuals that share the Earth will die in the next decade regardless of what anybody does, but I'm sure you're already well aware that species are defined by individuals, rather than the other way round. Even if the Red Squirrel Survival Trust (RSST) isn't:

'The red squirrel is at risk [singular (all = one)]. Without action it will become extinct in England within a decade. The Scottish, Welsh and Northern Irish red squirrel populations are also endangered. If we do not take urgent steps to address the issue, the red squirrel will disappear from the UK within a generation [guilt] …However, if we act now, we can protect our red squirrel population [faith]…The Red Squirrel Survival Trust receives no government funding – so we exist thanks to donations and the generosity of the public [penance].' [386]

And I must mention IFAW's whales again:

'Whales face more threats today' than at any time in history[guilt]…Your financial gift [penance] helps us to fight against commercial whaling, continue the non-lethal research of whales, protect endangered whale species and critical whale habitats, and promote responsible whale watching as a sustainable alternative to whaling [faith].' [387]

Can I mention IFAW's 'gift of hope' claim again as well?

> 'By taking direct action on IFAW campaigns [penance], you will be giving hope to thousands of animals around the world [faith] that suffer from cruelty [guilt], the risk of extinction [guilt], or need rescue from an emergency crisis [guilt].' [388]

I think faith is probably a horrendous understatement in this case though, because the belief that animals might *actually* receive hope because of your guilt fuelled penance surely requires madness as well as faith.

Whatever it involves you will never find redemption with these people. Firstly because things are just getting worse, which I will come back to in a moment, and secondly because their livelihoods are based on your donations now. Using their subjective beliefs they have generated a vast conservation economy and thus they will *not* stop. They will just move on to the next issue and the next and so on *ad infinitum*.

Perhaps peace-promising Harp Seal fanatic Rebecca Aldworth can explain:

> 'We have achieved a landmark victory for the seals! Today [5/5/2009], the European Parliament made history when it voted overwhelmingly to ban trade in seal products.' [389]

This was the issue that inspired her to bribe people with the chance to win a holiday by the way:

> 'We could not have won this victory without you. It is likely that when the seal hunt officially ends of 15 May, a quarter of a million seals will have been spared a horrible death. Now that the EU has banned its trade in seal products, countless more seals will live out their lives in peace from this year forward.' [390]

So far so good. Apart from that hideously subjective claim about 'peace' of course. There has to be a job-justifying caveat somewhere though:

'But...'

Here we go then:

> '...even as we celebrate this amazing victory, we must remain vigilant. Because there is every chance the sealing industry will develop new markets...Please join us as we finally put an end to the cruel slaughter of baby seals [guilt]. Become a monthly supporter of Humane Society International and help us keep the pressure on Canada [penance]...I know we can count on you to stay with us as we bring a final end to Canada's commercial seal slaughter [faith and penance].' [391]

Job justified.

I should say, again, that it's not cruel to kill baby seals per se, but that's not what I'm talking about. I'm saying that swiftly killing a Harp Seal is no worse than swiftly killing a chicken and much better than not killing the 500,000 dairy cows that are limping right now.

I'm also saying that this particular 'steward' won't stop making you feel guilty whatever happens. In fact, none of them will. For instance, IFAW has also issued a response to the European Parliament's ban on trading seal products:

> 'We now need your help more than ever to end this cruel and unnecessary hunt.' [392]

Apparently, victory has made things even *more* urgent and it should be perfectly obvious therefore that you will never find redemption.

Here's the WWF on recent increases in the the number of

African rhinos:

> 'Rhino conservation in Africa is going from strength to strength.' [393]

Wow, apart from coming at the expense of Saiga antelope populations, and apart from recent press releases warning that 'Poachers in Africa and Asia are killing an ever increasing number of rhinos'[394], that's great news. Can we relax now?

> '…there is still work to do.' [395]

Of course there is. How foolish of me:

> '…While taking stock and celebrating its many successes, the ARP [African Rhino Programme] is aware that much still needs to be done to secure a future for Africa's rhinos…Help us continue this success story [£££]…' [396]

Do I really need to point out that the future simply can't ever possibly be secure?

> 'These conservation battles must be won again and again.' [397]

And do I really need to point out that the WWF directly employs, and thus economically supports, nearly five and a half thousand people worldwide?

Here they are again on recent increases in Panda numbers:

> '…after years of decline, panda numbers are thought to be increasing.' [398]

Not quite as convincing as the Rhinoceros success I admit, but potentially great news nonetheless. Can we relax though?

'…there is still work to do.' [399]

Of course there is:

> 'The IUCN's red list classifies the panda as endangered,
> as its numbers remain low, despite the recent increase,
> and threats to its] survival remain…Help us continue
> this success story…' [400]

Do I really need to point out that 'the panda' is actually 'the pandas' and that they will all die whatever you do and that threats to the survival of the species will never go away?

The point is you will never find salvation. Never. The 'stewards' are fighting the endless change of a restless planet and that will never end. And besides, they're not even achieving anything anyway. We've been buying salvation for years but apart from some transient successes with a few superficially unique 'flagship' species the environment is still changing and that means even more guilt and even more penance, as the Harvard Business School has recognised:

> 'Most scientists agree that the earth is deteriorating at
> a faster rate than during the 1960's and 1970's, but it
> would be worse off had it not been for the tireless
> campaigning of environmental NGO's [non-govern-
> mental organisations].' [401]

Please note, things aren't just not better, they're getting worse and they're getting worse faster than they were before, i.e. environmental NGO's have achieved more than nothing without actually achieving anything at all.

The WWF linked *themselves* to this report by the way (via a link called *'Has the world become better off environmentally since* [WWF and Greenpeace] *were formed?'*[402]). I guess the most important part for them is another of the report's conclusions:

> 'WWF and Greenpeace create value by increasing the world's willingness-to-pay on environmental issues.' [403]

Which doesn't really need an awful lot of explaining does it.

Do you want to hear the UN's assessment of current environmental success:

> 'We are consuming more than nature can regenerate and we are producing wastes faster than Earths systems can process. Studies released in 2008 reinforce the message clearly: Human consumption of Earth's resources outstrips the planet's capacity to regenerate by about 30 per cent…As a result of continuing population growth and increasing material demands in many parts of the world, this ecological deficit is amplified each year.' [404]

The study they mention was the Living Planet Report (2008), by none other than…the WWF, who continue to believe they're magnificent:

> 'WWF has a history of delivering solid conservation results in many of the world's most biologically rich areas…Our goal is to build on past success to create lasting change that will allow us to live in harmony with our natural environment…We know our plans will work.' [405]

Do you want to have a look at some of the WWF's 'solid conservation results?

> 'Our climate change programme involves lobbying national governments around the world to reduce greenhouse gas emissions…As a very experienced and influential global organisation, WWF is able to exert great pressure – we're an instrumental force behind

the Kyoto Protocol, and have a strong track record of effecting change – a record we intend to not only keep but improve.' [406]

A 'strong track record of effecting change' remember, based on lobbying national governments to 'reduce green house gas emissions'. You would assume that would mean some sort of reduction somewhere. I wonder what the UN thinks:

> 'Since 2000, anthropogenic carbon dioxide emissions have been increasing four times faster than in the previous decade…These emissions are now 38 per cent above those in 1992, the year governments attending the Earth Summit, pledged to prevent dangerous climate change.' [407]

Pretty 'solid' isn't it, and yet they still manage to claim that they can protect 'the Polar Bear':

> 'Thank you. Your investment will help save a species and its habitat…We know our plans will work.' [408]

Can I mention their support for aboriginal polar bear hunting?

> 'WWF respects the rights of indigenous peoples to harvest marine mammals [including polar bears] in a responsible manner'[409]

And can I point out that every polar bear alive today will be dead by 2040 whatever they do, despite their own quite ludicrous beliefs?

> 'Your support will make the difference between life and death.' [410]

The WWF isn't trying to save 'the polar bear' on its own of

course, so it would be a little bit unfair to concentrate on them entirely, but the collective result is exactly the same. After 48 years of the WWF, 17 years of polar bears international, 38 years of Greenpeace, 106 years of Fauna and Flora International. 38 years of Friends of the Earth, 114 years of the Wildlife Conservation Society, 44 years of the IUCN/SSC Polar Bear Specialist Group etc. etc., here's the IUCN's simple assessment of the current population trend:

'Decreasing.' [411]

Good job everybody.

At least they haven't accidentally killed off their conservation target I suppose, unlike the Jaguar Conservation Team (JAGCT) of the Arizona Game and Fish Department (AZGFD).

(You will have to indulge me with this story. It's far too long, but it just gets more and more ridiculous and I just can't bring myself to delete it.)

Perhaps I should tell you a little bit about the expertise involved in the AZGFDJAGCT before I say too much more, or at least allow *them* to tell you anyway:

'A Scientific Advisory Committee advises JAGCT on its objectives in and approaches to jaguar conservation. The advisory committee includes several of the most renowned and well-published jaguar conservation experts in the world, as well as expert veterinarians and two scientists who have been working with borderlands jaguars for a decade or more.' [412]

And a little bit about the JAGCT goal:

'Protect and conserve Jaguars in Arizona and New Mexico.' [413]

And the AZFGD mantra:

'Managing today for wildlife tomorrow.' [414]

And then let them explain what good 'managers' they are:

> 18/2/09 - 'Jaguar conservation has just experienced an exciting development with the capture and collaring of the first wild jaguar in Arizona…The collared jaguar weighed in at 118 pounds with a thick and solid build. Field biologists' assessment shows the cat appeared to be healthy and hardy.' [415]

Please note, this cat was of 'thick and solid build' and appeared to be 'healthy and hardy', which means it must have been looking after itself pretty well.

> 28/2/09 – 'Updated tracking data indicates a change in the jaguars movement.' [416]

Sounds ominous.

> 1/3/09 – 'The cat's condition seems to be deteriorating.' [417]

Oh no. Perhaps this situation needs a bit more 'management'.

> 2/3/09: 'Macho B was euthanized after expert veterinarians at the Phoenix Zoo determined through blood tests and physical exam that the cat was in severe and unrecoverable kidney failure.' [418]

That's right, two weeks after being accidentally trapped by the AZFGD, and probably as a direct result of the sedatives he received during the capture, Macho B was dead. And that means what may well have been the 'last wild jaguar in Arizona'[419] was accidentally killed by people whose mission is to 'protect and conserve Jaguars in Arizona and New Mexico'.

To be fair to the AZGFD, he may also have been the oldest

jaguar in the world as well, but either way it's still a conservation disaster, as they humbly accept:

> 'We staunchly support the capture of the jaguar…' [420]

Well, some of them do anyway. Steve Spangle, U.S. Fish and Wildlife Service's Arizona field supervisor, clearly doesn't.
The AZFGD itself definitely does though:

> 'Macho B's death is upsetting and disappointing for everyone involved in wildlife conservation.' [421]

Although they have found some quite splendid ways of post-rationalising the situation as well. Here's Emil McCain, an active member of the Borderlands Jaguar Detection Project, for example:

> 'Before Macho B passed on he presented himself to the research and conservation efforts of an amazing collaborative Arizona/New Mexico Jaguar Conservation Team.' [422]

Would that be the 'amazing' collaborative Arizona/New Mexico Jaguar Conservation Team that just managed to kill him Emil?
He continues:

> 'He gave us valuable DNA…' [423]

He what?

> 'He gave us valuable DNA which will give us genetic information about his origin, his relatedness to other Jaguars [he could be a different 'species' remember], and thus the viability of borderlands jaguars. His death is terribly sad. But it is now up to us to cherish and learn from Macho B's gift…' [424]

His what? His Gift?

> '…and we must work hard for the continued presence
> of his kind in our wild country.' [425]

Emil, please. There's a good chance he was the last one and
you and your AZFGDJAGCT friends just killed him.

> 'Despite the tragic outcome, the Jaguar Conservation
> Team and its co-operators have pulled off an amazing
> feat.' [426]

I'm sorry. Did he just call killing the last Jaguar in Arizona a
'tragic outcome' and an 'amazing feat' at the same time?
     Yes. He did. And Larry Voyles, the Director of the Arizona
Game and Fish Department, has also found a silver lining:

> 'We don't know if Macho B is [was] the last wild jaguar
> in Arizona, but…[he] is a symbol of hope for our open
> spaces and the connectivity of our habitats.' [427]

Yes, you read it correctly: the dead jaguar *his* organisation just
killed, is a symbol of hope. And Linda Valdez of the Arizona
Republic newspaper agrees:

> 'Yet the passions surrounding this animal's death are
> the good news in this story. These feelings show how
> far Arizona and the nation have advanced from the
> days when predators were systematically killed. This
> cat didn't die because of efforts to eradicate his breed.
> He died after heroic efforts to save him.' [428]

But Linda, he didn't need 'heroic efforts to save him'. Well, he
did, but only from the Arizona jaguar 'stewards'.
     Incidentally, here's what the AZFGD thinks of animals
that *aren't* old jaguars:

'Arizona is well known throughout the country and abroad for its quality big game hunting opportunities.' [429]

You guessed it. You can shoot everything else for fun and, despite Linda's conclusions about how far 'the nation has advanced from the days when predators were systematically killed', that includes predators:

'Game species include…mountain lion.' [430]

I wonder why Jaguars warrant their own conservation team, while mountain lions just warrant 'considerable skill executing…the kill'[431]:

'Jaguars are the only cat in North America that roars.' [432]

I see. Is that it?

'Jaguars are easily distinguished from mountain lions (pumas)…by their pronounced spots.' [433]

Brilliant. Jaguars roar and they're spotty, while mountain lions don't and they aren't. What perfect morphological racism.

Anyway, despite killing what may have been the last Jaguar in Arizona, the AZFGDJAGCT still manages to maintain a high opinion of itself:

'JAGCT…efforts and those of colleagues in Mexico are helping create a more promising future for the jaguar in the U.S.-Mexico borderlands.' [434]

But that just goes with the territory, because all 'stewards' have a high opinion of themselves, regardless of what they have managed to achieve (or accidentally kill). Greenpeace is currently 'celebrating 10 years of saving the Amazon Rainforest'[435] for example, but how much of the Amazon

Rainforest have they actually saved? Like them I will focus on the Brazilian Amazon, but unlike them I will base my celebrations on the facts, as reported by the Food and Agriculture department of the United Nations (FAO). According to the FAO then, in 1999 (the beginning of the 10 year period Greenpeace is currently celebrating) there were 495,894,400 Ha of forest in Brazil. And in 2007 (the most recent data available) there were 471,492,000 Ha. That's a net loss of nearly 2.5 *million* Ha in 8 celebration-justifying years[436].

To be fair to Greenpeace, the FAO figures do refer to the whole of Brazil, rather than just the Brazilian Amazon, but according to the WWF focused Amazonian statistics don't really justify a celebration either:

> 'Despite ongoing conservation efforts, the Amazon is losing on average 27,000km2 of forest cover each year.' [437]

In fact, deforestation in general is another example of how ineffective all of these organisations have been in delivering the salvation they promised to relieve the guilt they generated. Despite collecting and spending billions of pounds/euros/dollars of guilt tax in the last 19 years (the WWF alone has spent over $1 billion since 1985), global forest cover has dropped by nearly 1.4 *million* km².

As always though, the past is the past and the future is shrouded in hope and potential:

> 'Big problems need big solutions and collective action: Greenpeace is ready for both.' [438]

I'm not entirely sure what they have been doing while small problems requiring small solutions have become big problems needing big solutions, but after 38 years of achieving more than nothing without actually achieving anything, Greenpeace is finally ready to make a difference. At least that's the belief they will hide behind when it comes to extracting more penance in the face of continued failure, and the IUCN is the same:

> 'With our reputation for generating and disseminating sound scientific knowledge, our diverse structure and credibility that allows us to convene a range of stakeholders around key problems and our local to global reach, IUCN can play a catalytic role in this renewed global effort.' [439]

I'm not exactly sure what the world's 'largest professional global conservation network' has been doing for the last 61 years (other than writing a lot of long reports), but the past is the past and they're now ready to catalyse a new global effort.

They're an organisation which also maintains a very high opinion of itself by the way, even though they openly accept that 'despite all of the activity in the environmental movement during the latter part of the 20th century, we have come little closer to answering the fundamental question of how to deliver sustainability'[440], and honestly describe current conservation efforts as, at the very best, '…walking north on a southbound train'[441]:

> 'For sixty years, IUCN has led the development of conservation science and knowledge, and brought together governments, NGOs, scientists, companies and community organizations to help the world make better conservation and development decisions.' [442]

Like Greenpeace they have even found reason to celebrate themselves as well:

> '…IUCN celebrates its 60th anniversary, and marks six decades of global conservation achievement…' [443]

But what exactly have they achieved?

> '...the challenges ahead are bigger than anything we have ever faced before...' [444]

Brilliant.

What the IUCN, and all the other 'stewards', seem to have forgotten is that population size is the real problem, and it's still growing. There will be 32% (2.2 *billion*) more people by 2050 remember. In the words of Sir David Attenborough:

> '...behind every threat is the frightening explosion in human numbers...I've never seen a problem that wouldn't be easier to solve with fewer people, or harder, and ultimately impossible, with more.' [445]

Even Professor Lovelock agrees:

> '...the root cause is too many people...' [446]

The 'stewards' will not give up the faith however and biodiversity is another conservation priority that has been used to elicit vast quantities of financial penance. Here is the WWF to explain how much better everybody should feel as a result:

> 'The Living Planet Index of global biodiversity, as measured by populations of 1,686 vertebrate species across all regions of the world, has declined by nearly 30 per cent over just the past 35 years.' [447]

'30 per cent over just the past 35 years'. Should I point out that the WWF has been campaigning for 48 years? Which, incidentally, is *exactly* the same amount of time it has taken for humanities ecological 'footprint' to double as well:

> 'Human demand on the biosphere more than doubled

during the period 1961 to 2005.' [448]

Basically, I can't keep up with the amount of evidence I can find about how fast everything is continuing to change. Apart from a few transient increases in the numbers of some popular animals almost every parameter you can think of shows that these organisations haven't delivered on the promises they have made about the salvation we can expect from the penance we have given, and the only way they can continue asking for money is by extolling the virtue of belief without evidence, and faith without fact. In the words of the IUCN, again:

> 'Hope is about defying the odds. It is the faith that things will work out whatever the odds.' [449]

Have I justified the original question enough yet?

> What do you call something that uses subjective belief to promote guilt while simultaneously offering salvation through penance and faith?

Actually, it should say:

> What do you call something that uses subjective belief in *secular immortality* to promote guilt while simultaneously offering salvation through penance and faith?

I should probably justify 'secular immortality' as well then. That's easy though. Firstly because immortality is the whole point of not letting things change (i.e. conservation), and secondly because the disassociation of individuals from species (i.e. all = one) has replaced the reality of death with the prospect of extinction and allowed the 'stewards' to create immortal species they can condemn us for threatening. By convincing everybody that 'extinction is forever' they have conveniently bypassed the fact that 'death is also forever' and created an entire portfolio of immortal entities, like 'the tiger', 'the whale', 'the Orang-utan', 'the polar bear', 'the panda', 'the red squirrel', so that they can generate guilt and offer all mortal sinners a divine path to salvation. They're even trying to split them into more immortal species to generate even more guilt and thus even more cash.

Actually, the question *still* isn't complete, because there's one last factor to take into consideration:

> What do you call something that 'worships a god' at the same time as using subjective belief in secular immortality to promote guilt while simultaneously offering salvation through penance and faith?

'THE HUMAN'

Who's the new god then?

That's easy as well, because that's the entire point of making us all 'stewards'. By giving *Homo sapiens* responsibility for all Life on Earth the 'stewards' have created 'the human' and thus 'the Steward' and gone one step further than all previous attempts at narcissistic self-reverence. They have shaken off the restraints of supernatural superstition and turned us into our own farmer, son and holy steward. Now it's us that has the power to kill and save planets. Now it's us that has the power to grant species and ecosystem immortality. Now it's us that is lord of all we survey. Because of conservation we don't need God anymore; we *are* God:

> And Man said: Let us make a God of ourselves, after our likeness: and let us have stewardship over the fish of the sea, and over the fowl of the air, and over the cattle, and over all the Earth, and over every creeping thing that creepeth upon the Earth.

So, what do you call something that worships *itself* as a

merciful planetary god, at the same time as using subjective belief in the immortality of superficial appearance, to promote guilt about organic beings that are all going to die anyway, while simultaneously offering salvation from that guilt through financial penance, based entirely on claims that can't be checked and which totally ignore 3.5 *billion* years of proven success?

I don't know about you, but I would call it a religion.

Chapter 11

DESTINY

I'm still not a big fan of grey. I still prefer to see the world in black and white and I still think it's worth remembering that every aspect of every conflict can often be divided into *two* opposite and mutually exclusive possibilities. For instance, when we consider the Earth's cosmic significance, there are still just two sides to the debate: one, the Earth *is* the centre of the Solar System, or two, it isn't. And if we combine these possibilities with a bit of evidence, there's really just one *correct* side to the debate: the Earth is *not* the centre of the Solar System. Similarly, when we consider *our* cosmic significance there are also still just two sides to the debate: one, we *are* the centre of the Universe, or two, we're *not*. And, similarly again, using evidence there's really just one *correct* side to this debate as well: we're *not* the centre of the Universe. And the destiny of species is no different. When we consider the relations of the present to the past inhabitants of this world, there are really only two possible sides to the debate: one, species *are* transient morphological entities that will all disappear, or two, they're *not*.

But which is it? Is the Polar Bear a sacred evolutionary crescendo, or a possible evolutionary link? Is the tiger an iconic evolutionary summit, or an expendable evolutionary pathway? 'We see beautiful adaptations everywhere and in every part of the organic world'[Darwin], but they definitely haven't been created 'for the sake of beauty, to delight man or the creator'[Darwin] so where are they going and how will they get there?

What is the true *destiny* of species?

'Judging from the past, we may safely infer that not

one living species will transmit its unaltered likeness to a distant futurity. And of the species now living very few will transmit progeny of any kind to a far distant futurity; for the manner in which all organic beings are grouped, shows that the greater number of species in each genus, and all the species in many genera, have left no descendants, but have become utterly extinct.'[Darwin]

As usual Charles Darwin understood it all along. We may not be able to predict the specific details, but that doesn't matter, because we don't need to. To explain the result all we need is the process and thanks to Charles Darwin natural selection is a blindingly obvious progression that has been changing species for 3.5 billion years. And it will continue to change them in the future. As long as Planet Earth continues to support Life, the 'preservation of favoured races'[Darwin] and the elimination of unfavoured races will continue to select or eliminate until everything is either different or dead:

> 'It is interesting to contemplate a tangled bank, clothed with many plants of many kinds, with birds singing on the bushes, with various insects flitting about, and with worms crawling through the damp earth, and to reflect that theses elaborately constructed forms, so different from each other, and dependent upon each other in so complex a manner'[Darwin], will all change or die, whether the 'stewards' like it or not.

It doesn't matter whether they're domestic, or captive, or wild, or human. If 3.5 *billion* years of evolutionary divergence continues into the future then every species will change or die.

Every single one.

Why? Because that's the whole point. Evolution isn't about resisting change, it's about embracing it. It's not about preserving species, it's about preserving Life. Species and individuals are just transient vehicles through the prevailing conditions. They're soldiers of Life, pitched against each other and themselves in a dynamic and ruthless battle for survival. Just as it was 150 years ago:

> 'Looking to the future, we can predict that the groups of organic beings which are now large and triumphant, and which are least broken up, that is, which have as yet suffered least extinction, will, for a long period, continue to increase. But which groups will ultimately prevail, no man can predict; for we know that many groups, formerly most extensively developed, have now become extinct.'[Darwin]

We may have forgotten, but Charles Darwin was well aware that 'the extinction of old forms is the almost inevitable consequence of the production of new forms' and his Tree of Life was based on death and extinction (objective survival), not peace and love (sentimental and self-centered conservation):

> 'At each period of growth all the growing twigs have tried to branch out on all sides, and to overtop and kill the surrounding twigs and branches, in the same manner as species and groups of species have at all times overmastered other species in the great battle for life …Of the many twigs which flourished when the tree was a mere bush, only two or three, now grown into great branches, yet survive and bear the other branches; so with the species which lived during long-past geological periods, very few have left living and modified descendants…As buds give rise by growth to fresh buds, and these, if vigorous, branch out and overtop on all sides many a feebler branch, so by generation I believe it has been with the great Tree of Life, which fills with its dead and broken branches the crust of the earth, and covers the surface with its ever branching and beautiful ramifications.'

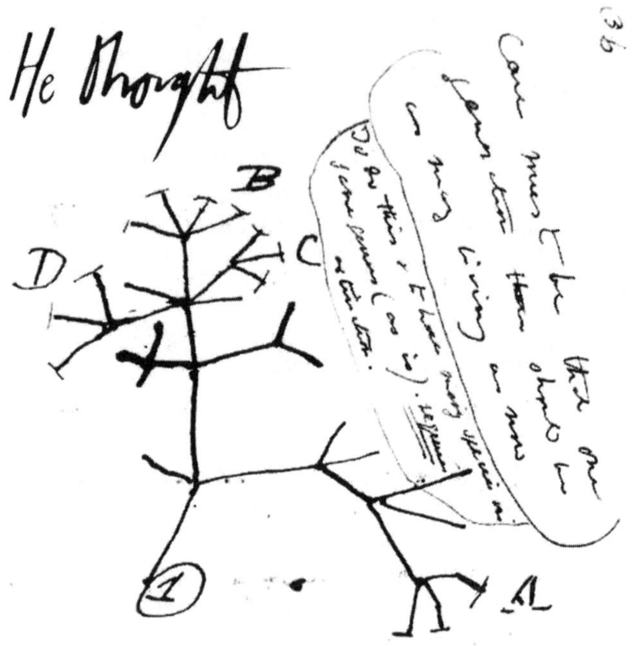

Charles Darwin's Tree of Life, as ignored by everybody who thinks evolution doesn't involve an awful lot of death and extinction (the capped branches).

In reality, rolling diversity is the *only* way to cope with changing conditions of life. It's the only way of coping with Milankovitch cycles, or plate tectonics, or global chemistry, or better predators, or hominid pyromania, or anything. There's no room for com-passionate conservation and sympathetic preservation and for Charles Darwin there never was:

'On the theory of natural selection, the extinction of old forms and the production of new and improved forms are intimately connected together.'

Everything changes and that means extinction is and always

has been a vital part of evolution, regardless of rate. It's essential and that's why conservation is *not* natural history. To really see nature we can't be emotional and *subjective*. We must be detached and *objective*. We must ignore 'the blindness of preconceived opinion' and embrace the vision of unconceived truth, because Life isn't about equality and 'love', it's about prejudice and survival. It's about the elimination of unfavoured races in the struggle for life:

> 'Thus, as it seems to me, the manner in which single species and whole groups of species become extinct accords well with the theory of natural selection. We need not marvel at extinction; if we must marvel, let it be at our own presumption in imagining for a moment that we understand the many complex contingencies on which the existence of many species depends.'[Darwin]

Nature isn't a dainty little lamb, or a frail and sickly baby mouse; it's a ruthless battle-scarred survival dragon; that will stand and face the wind of change with steel in its belly and fire in its heart. It's a merciless general, commanding a vast army of expendable soldiers, and, long after *Homo sapiens* have gone, it will continue to fight. Until Planet Earth can support the fight no more.

And 'there is grandeur in this view of Life'[Darwin].

FURTHER READING

ON

THE ORIGIN OF SPECIES

BY MEANS OF NATURAL SELECTION,

OR THE

PRESERVATION OF FAVOURED RACES IN THE STRUGGLE
FOR LIFE.

BY CHARLES DARWIN, M.A.,

FELLOW OF THE ROYAL, GEOLOGICAL, LINNÆAN, ETC., SOCIETIES;
AUTHOR OF 'JOURNAL OF RESEARCHES DURING H. M. S. BEAGLE'S VOYAGE
ROUND THE WORLD.'

LONDON:

JOHN MURRAY, ALBEMARLE STREET.

1859.

NOTES

References are listed with internet links where available. Please note, the internet is a changing world so some links will inevitably disappear. To address this as much as possible I have included the most recent date of retrieval (and taken copies of most, if not all quoted internet pages).

[1] Concise Oxford English Dictionary
http://www.askoxford.com/dictionaries/compact_oed

[2] Concise Oxford English Dictionary.

[3] Concise Oxford English Dictionary.

[4] Concise Oxford English Dictionary.

[5] IMAGE CREDIT- Pleple2000. Permission is granted to copy, distribute and/or modify this document under the terms of the GNU Free Documentation License, Version 1.2 or any later version published by the Free Software Foundation; with no Invariant Sections, no Front-Cover Texts, and no Back-Cover Texts.
http://commons.wikimedia.org/wiki/File:Buldog_angielski_000pl.jpg#file

[6] Bulldogsworld.com (31/08/2009).
http://www.bulldogsworld.com/index.html

[7] Bulldogsworld.com (31/08/2009).
http://www.bulldogsworld.com/ai.html

[8] Bulldogsworld.com (31/08/2009).
http://www.bulldogsworld.com/index.html

[9] Bulldogsworld.com (31/08/2009).
http://www.bulldogsworld.com/BREEDINGWHELP.html

[10] IMAGE SOURCE - This image or recording is the work of an U.S. Fish and Wildlife Service employee, taken or made during the course of an employee's official duties. As a work of the U.S. federal government, the image is in the public domain. For more information, see the Fish and Wildlife Service copyright policy.
http://commons.wikimedia.org/wiki/File:Canis_lupus_laying.jpg

[11] IMAGE CREDIT- Pleple2000. Permission is granted to copy, distribute and/or modify this document under the terms of the GNU Free Documentation License, Version 1.2 or any later version published by the Free Software Foundation; with no Invariant

Sections, no Front-Cover Texts, and no Back-Cover Texts. http://commons.wikimedia.org/wiki/File:Buldog_angielski_000pl.jpg#file

[12] IMAGE SOURCE - This image or recording is the work of an U.S. Fish and Wildlife Service employee, taken or made during the course of an employee's official duties. As a work of the U.S. federal government, the image is in the public domain. For more information, see the Fish and Wildlife Service copyright policy. http://commons.wikimedia.org/wiki/File:Canis_lupus_laying.jpg

[13] IMAGE CREDIT- Pleple2000. Permission is granted to copy, distribute and/or modify this document under the terms of the GNU Free Documentation License, Version 1.2 or any later version published by the Free Software Foundation; with no Invariant Sections, no Front-Cover Texts, and no Back-Cover Texts. http://commons.wikimedia.org/wiki/File:Peki%C5%84czyk_651.jpg

[14] Chris Lawrence. Balancing breed standards and welfare. British Small Animal Veterinary Association (BSAVA) Congress Times (2009).

[15] Bradley Viner. Isn't it about time pedigree dogs had a chance to bit back at their critics? Veterinary Times (2/3/2009).

[16] Crufts (31/08/2009). http://www.crufts.org.uk/

[17] Kennel Club Breed Standard (31/08/2009). http://www.thekennelclub.org.uk/item/36

[18] The Rhodesian Ridgeback Club of Great Britain (31/08/2009). http://rhodesianridgebacks.org/

[19] Kennel Club Breed Standard (31/08/2009). http://www.thekennelclub.org.uk/item/198

[20] Kennel Club Breed Standard (31/08/2009). http://www.thekennelclub.org.uk/item/199

[21] Kennel Club Breed Standard (31/08/2009). http://www.thekennelclub.org.uk/cgi-bin/item.cgi?id=190

[22] IMAGE CREDIT – Alex Brown. Reprodcued under a Creative Commons Attribution license (CC-BY): Attibution 2.0 Generic. http://www.flickr.com/photos/alexbrn/3140892215/

[23] Bradley Viner. Isn't it about time pedigree dogs had a chance to bit back at their critics. Veterinary Times (2/3/2009).

[24] Oli Viner. Whispering my pedigree passions. Veterinary Times (20/4/2009).

[25] FrenchBulldogZ (31/08/2009).
http://frenchbulldogz.org/purchasing/health.htm

[26] FrenchBulldogZ (31/08/2009).
http://www.frenchbulldogz.org/learn/breedingrisks.htm

[27] Bradley Viner. Isn't it about time pedigree dogs had a chance to bit back at their critics. Veterinary Times (2/3/2009).

[28] Caroline Kisco. Who's in the doghouse. Sunday Times newspaper (1/3/2009).

[29] David Coffey. Letters to the Editor. Veterinary Times (2009).

[30] The International Cat Association (4/9/2009).
http://tica.org/public/breeds/mk/intro.php

[31] University of Oxford Animal Behaviour Research Group. A Review of the Welfare of Zoo Elephants in Europe (2002) (commissioned by the RSPCA).
http://www.rspca.org.uk/servlet/ContentServer?pagename=RSPCA/RSPCARedirect&pg=WildlifeReportsandResources

[32] University of Oxford Animal Behaviour Research Group. A Review of the Welfare of Zoo Elephants in Europe (2002) (commissioned by the RSPCA).
http://www.rspca.org.uk/servlet/ContentServer?pagename=RSPCA/RSPCARedirect&pg=WildlifeReportsandResources

[33] Hindu Forum of Britain (3/9/2009).
http://www.hfb.org.uk/default.aspx?sID=762&cID=219&ctID=43&lID=0

[34] Hindu Forum of Britain (3/9/2009).
http://www.hfb.org.uk/default.aspx?sID=762&cID=219&ctID=43&lID=0

[35] Hindu Forum of Britain (3/9/2009).
http://www.hfb.org.uk/default.aspx?sID=762&cID=219&ctID=43&lID=0

[36] Hindu Forum of Britain (3/9/2009).
http://www.hfb.org.uk/default.aspx?sID=762&cID=219&ctID=43&lID=0

[37] Hindu Forum of Britain (3/9/2009).
http://www.hinduforum.org/fileserver.aspx?oID=347&lID=0

[38] Hindu Forum of Britain (3/9/2009).
http://www.hfb.org.uk/default.aspx?sID=762&cID=219&ctID=43&lID=0

[39] Stuart Coyle(3/9/2009). http://www.hinduforum.org/fileserver.aspx?oID=347&lID=0

[40] Hindu Forum of Britain (3/9/2009). http://www.hfb.org.uk/default.aspx?sID=762&cID=219&ctID=43&lID=0

[41] Bhaktividenta Manor. http://www.krishnatemple.com (link removed).

[42] Bhaktividenta Manor. http://www.krishnatemple.com (link removed).

[43] Gauri Das (3/9/2009). http://www.dandavats.com/?p=4938#more-4938

[44] Bhaktividenta Manor (3/9/2009). Official Statement from Bhaktividenta Manor. http://www.dandavats.com/?p=6092

[45] Bhaktividenta Manor (3/9/2009). Official Statement from Bhaktividenta Manor. http://www.dandavats.com/?p=6092

[46] DEFRA. Codes of Recommendations for the Welfare of Cattle. http://www.defra.gov.uk/animalh/welfare/farmed/cattle/booklets/cattcode.pdf

[47] Hindu Forum of Britain (3/9/2009). http://www.hfb.org.uk/default.aspx?sID=762&cID=219&ctID=43&lID=0

[48] Bhaktividenta Manor (3/9/2009). http://www.justiceforgangotri.org/2008/03/30/press-release-hindus-to-sue-rspca-for-death-of-the-sacred-cow-gangotri/

[49] Hindu Forum of Britain (3/9/2009). http://www.hinduforum.org/Default.aspx?sID=45&cID=348&ctID=43&lID=0

[50] Bhaktividenta Manor (3/9/2009). http://www.justiceforgangotri.org/2008/02/14/the-scattering-of-gangotris-ashes-gauri-das-reports-from-india/

[51] Mahatma Ghandi. Young India (1926).

[52] IMAGE CREDIT – Jann-Erik Finnberg. Reprodcued under a Creative Commons Attribution license (CC-BY): Attibution 2.0 Generic. http://www.flickr.com/photos/wheany/3226834064/

[53] Carl Padgett. For better or worse? BVA AWF forum considers breeding, and electric shock collars. Veterinary Record (7/6/2008).

[54] Emma Milne. For better or worse? BVA AWF forum considers breeding, and electric shock collars. Veterinary Record (7/6/2008).

[55] Martin Atkinson. Do welfare considerations equal stress? Veterinary Times (10/8/2009).

[56] Chris Lawrence. Balancing pedigree dog breed standards and animal welfare. Veterinary Record (18/4/2009).

[57] Karen Humm Balancing pedigree dog breed standards and animal welfare. Veterinary Record (18/4/2009).

[58] David Coffey. What are the possible causes? Veterinary Practice magazine (5/2008).

[59] David Coffey. Letters to the Editor. Veterinary Times (31/8/2009).

[60] David Bee. Letters to the Editor. Veterinary Times (3/8/2009).

[61] David Coffey. What are the possible causes? Veterinary Practice (05/2008).

[62] RCVS Guide to Professional Conduct, Section 1A.
http://www.rcvs.org.uk

[63] RCVS Guide to Professional Conduct, Section 1F.
http://www.rcvs.org.uk

[64] RCVS Guide to Professional Conduct, Section 1B.
http://www.rcvs.org.uk

[65] RCVS Guide to Professional Conduct, Section 2C.
http://www.rcvs.org.uk

[66] European Food Safety Authority. Scientific Opinion on the overall effects of farming systems on dairy cow welfare and disease 2009.
http://www.efsa.europa.eu/EFSA/efsa_locale-1178620753812_1211902630995.htm

[67] Bulldogsworld.com (31/08/2009).
http://www.bulldogsworld.com/index.html

[68] RCVS Guide to Professional Conduct, Section 2A.
http://www.rcvs.org.uk

[69] RCVS Guide to Professional Conduct, Section 1C.
http://www.rcvs.org.uk

[70] Current attitudes of cattle practitioners to pain and the use of analgesics in cattle. Veterinary Record (11/11/2006).

[71] Philip Lowe. Unlocking Potential. A report on veterinary expertise in food animal production.
http://www.defra.gov.uk/animalh/ahws/vservices/pdf/lowe-vets090806.pdf

[72] RCVS . The UK Veterinary Profession in 2006.
http://www.rcvs.org.uk/Shared_ASP_Files/UploadedFiles/97D3DAF
0-9567-4B2F-9972-B0A86CFB13C7_surveyprofession2006.pdf

[73] Lack of Appraisals due to high workload, vets claim. Veterinary
Times (22/6/2009).

[74] RCVS Guide to Professional Conduct, Annex P.
http://www.rcvs.org.uk

[75] RCVS Guide to Professional Conduct, Annex P.
http://www.rcvs.org.uk

[76] RCVS Guide to Professional Conduct, Section 1A.
http://www.rcvs.org.uk

[77] RCVS Guide to Professional Conduct, Annex P.
http://www.rcvs.org.uk

[78] RCVS Guide to Professional Conduct, Annex P.
http://www.rcvs.org.uk

[79] Frank Busch. Considering ethics and anaesthetics. Veterinary
Times (8/12/2008).

[80] RCVS. http://www.rcvs.org.uk/AboutRCVS

[81] Food Standards Agency. The National Diet & Nutrition Survey:
adults aged 19 to 64 years. Summary Report (Volume 5). Published
2004.

[82] DEFRA Report, questionnaire and data tables following Survey of
Public Attitudes and Behaviours toward the Environment: 2007.

[83] Food Standards Agency – Public Attitudes to Food Issues (2009).

[84] United Nations Food and Agriculture Organisation. Torry
Advisory Note No. 11, Handling Inshore Fish. United Nations Food
and Agriculture Organisation.

[85] United Nations Food and Agriculture Organisation. Torry
Advisory Note No. 34 (revised), Freezing at Sea: Advice to the Crew..

[86] United Nations Food and Agriculture Organisation. Torry
Advisory Note No. 50, Some Notes on fish Handling and Processing.

[87] V. D. Vis and Kesten. 1996. Killing of fishes; literature-study and
practice- observations (field research) report number C 037/96,
1996 RIVO DLO.

[88] Deepwater Fishing Along The North Atlantic Slope. Fishing Trials
Onboard MFV "MARANATHA III". Seafish (www.seafish.org).

[89] PETA (1/9/2009). http://www.peta.org/sea_kittens/about.asp

[90] PETA (1/9/2009). http://www.peta.org/sea_kittens/book.asp

[91] Vegetarian Society (1/9/2009).
http://www.vegsoc.org/Info/definitions.html

[92] DairyUK (1/9/2009).
http://www.dairyuk.org/index.php?option=com_content&task=vie
w&id=206&Itemid=286

[93] DairyUK (1/9/2009). http://www.proudofdairy.com/

[94] DairyCo. http://www.dairyco.org.uk (link removed, although I do
have a copy).

[95] DairyCo (1/9/2009). http://www.dairyco.org.uk/about-us/what-
is-dairyco/dairyco's-mission.aspx

[96] DairyCo (1/9/2009). http://www.dairyco.org.uk/about-us/what-is-
dairyco/how-is-dairyco-funded.aspx

[97] DairyCo. http://www.dairyco.org.uk (link removed, although I do
have a copy).

[98] DairyCo. http://www.dairyco.org.uk (link removed, although I do
have a copy).

[99] DairyUK (1/9/2009). The White Paper 2009.
http://www.dairyuk.org/

[100] DairyCo (1/9/2009). http://www.dairyco.org.uk/farming-info-
centre/health--welfare/mobility/dairyco-mobility-score.aspx

[101] DEFRA.
http://www.defra.gov.uk/animalh/welfare/farmed/advice/mobility.pdf

[102] DairyCo. Factors Affecting Milk Supply 2009.
http://www.dairyco.org.uk/library/market-information/milk-
supply/factors-affecting-milk-supply-2009.aspx

[103] National Animal Disease Information Service (NADIS)
(1/9/2009)..
http://www.nadis.org.uk/DiseasesCattle/Endometritis/ENDOME_1.
htm

[104] Dairy Herd Health and Productivity Service (DHHPS) run by
Edinburgh University (1/9/2009).
http://www.vet.ed.ac.uk/DHHPS/documents/MilkFeverControlPDF.pdf

[105] DairyUK (1/9/2009).
http://www.dairyuk.org/index.php?option=com_content&task=vie
w&id=206&Itemid=286

[106] Hayley Campbell-Gibbons (1/9/2009).
http://www.nfuonline.com/x38316.xml

[107] European Food Safety Authority. Scientific Opinion on the overall effects of farming systems on dairy cow welfare and disease 2009. http://www.efsa.europa.eu/EFSA/efsa_locale-1178620753812_1211902630995.htm

[108] DairyCo (1/9/2009). http://www.dairyco.org.uk/farming-info-centre/health--welfare/mobility.aspx

[109] Hayley Campbell-Gibbons (1/9/2009). http://www.nfuonline.com/x38316.xml

[110] European Food Safety Authority. Scientific Opinion on the overall effects of farming systems on dairy cow welfare and disease 2009. http://www.efsa.europa.eu/EFSA/efsa_locale-1178620753812_1211902630995.htm

[111] DairyCo. Strategies to reduce culling rates 2008. http://www.dairyco.org.uk/library/farming-info-centre/health--welfare/strategies-to-reduce-culling-rates-(d).aspx

[112] Roger Evans. Attending to detail, keeping clean. Veterinary Times (24/8/2009).

[113] European Food Safety Authority. Scientific Opinion on the overall effects of farming systems on dairy cow welfare and disease 2009. http://www.efsa.europa.eu/EFSA/efsa_locale-1178620753812_1211902630995.htm

[114] Hayley Campbell-Gibbons (1/9/2009). http://www.nfuonline.com/x38316.xml

[115] Red Tractor (1/9/2009). http://www.redtractor.org.uk/site/REDT/Templates/GeneralStandards.aspx?pageid=32&cc=GB

[116] John Blackwell. Farmers Weekly (1/9/2009). http://www.fwi.co.uk/Articles/2009/05/13/115606/dairy-industry-defends-itself-against-rspca-welfare.html

[117] Keith Cutler. Veterinary Times (5/2009).

[118] Hugh Fearnley-Whittingstall (1/9/2009). http://www.channel4.com/food/on-tv/river-cottage/hugh-fearnley-whittingstall-fish-interview-07-11-08_p_2.html

[119] NOAA. http://www.nmfs.noaa.gov/fishwatch/species/walleye_pollock.htm (page since updated).

[120] Richard Dawkins. The Selfish Gene (1976).

[121] IMAGE CREDIT - Toivo Toivanen & Tiina Toppila. This file is in the public domain, because the author provided the image "for unlimited free use" on the source site.
http://commons.wikimedia.org/wiki/File:Kurre6.jpg

[122] IMAGE CREDIT – Dave-F. Reprodcued under a Creative Commons Attribution license (CC-BY): Attibution 2.0 Generic.
http://www.flickr.com/photos/frield/2252687794/

[123] Red Squirrel Survival Trust (1/9/2009).
http://www.rsst.org.uk/index.php?option=com_content&view=article&id=75&Itemid=107

[124] Red Squirrel Survival Trust (1/9/2009).
http://www.rsst.org.uk/index.php?option=com_content&view=article&id=49&Itemid=88

[125] Red Squirrel Survival Trust (1/9/2009).
http://www.rsst.org.uk/index.php?option=com_content&view=article&id=49&Itemid=88

[126] Red Squirrel Survival Trust (1/9/2009).
http://www.rsst.org.uk/index.php?option=com_content&view=article&id=49&Itemid=88

[127] European Squirrel Initiative (1/9/2009).
http://www.europeansquirrelinitiative.org/the_threat.html

[128] European Squirrel Initiative (1/9/2009).
http://www.europeansquirrelinitiative.org/the_threat.html

[129] Red Squirrel Protection Partnership (1/9/2009).
http://www.rspp.org.uk/id22.html

[130] WWF (1/9/2009).
http://www.panda.org/about_our_earth/blue_planet/problems/shipping/alien_invaders/

[131] WWF (1/9/2009).
http://www.panda.org/about_our_earth/blue_planet/problems/shipping/alien_invaders/

[132] US National Science Foundation (1/9/2009).
http://www.nsf.gov/news/special_reports/jellyfish/textonly/swarms.jsp

[133] The jellyfish joyride: causes, consequences and management responses to a more gelatinous future: Trends in Ecology & Evolution (6/2009).

[134] The jellyfish joyride: causes, consequences and management responses to a more gelatinous future: Trends in Ecology & Evolution (6/2009).

[135] The Australian Museum (1/9/2009).
http://australianmuseum.net.au/Cane-Toad

[136] The University of Queensland (1/9/2009).
http://www.imb.uq.edu.au/index.html?page=48437

[137] IUCN (1/9/2009). http://www.iucnredlist.org/details/41065/0

[138] ZSL (1/9/2009). http://www.zsl.org/conservation/carnivores-and-people/zoo-tigers-and-tiger-conservation,359,AR.html

[139] IUCN (1/9/2009). http://www.iucnredlist.org/details/41585/0

[140] Songbird Survival Trust (1/9/2009). http://www.songbird-survival.org.uk/predators/magpies/

[141] Chris Packham (1/9/2009).
http://www.guardian.co.uk/theguardian/2009/apr/20/magpies-protect-cull-songbird-survival

[142] RSPB (1/9/2009).
http://www.rspb.org.uk/advice/gardening/unwantedvisitors/predators_and_prey/in_conclusion.asp

[143] RSPB (1/9/2009).
http://www.rspb.org.uk/advice/gardening/unwantedvisitors/predators_and_prey/controlling_predators.asp

[144] David Hoccom (1/9/2009). The Times.
http://www.timesonline.co.uk/tol/news/environment/article6115155.ece

[145] Rebecca Aldworth. HSI campaign email.

[146] Rebecca Aldworth. HSI campaign email.

[147] Rebecca Aldworth. HSI campaign email.

[148] IFAW (1/9/2009).
http://www.ifaw.org/ifaw_international/join_campaigns/protecting_whales_around_the_world/index.php

[149] IWC. Annual Report of the International Whaling Commission 2005. Annex F - Report of the Working Group on Whale Killing Methods and Associated Welfare Issues.

[150] IWC. Annual Report of the International Whaling Commission 2003. Annex E - Report of the Workshop on Whale Killing Methods and Associated Welfare Issues.

[151] NFU (1/9/2009). http://www.greatbritishchicken.co.uk/meet-the-farmer/fact-chicken.pdf

[152] Red Tractor Assured Chicken Production standards 2008-2009.
http://www.assuredchicken.org.uk/chicken/scheme/standard.eb

153 DEFRA. Codes of Recommendations for the Welfare of Livestock – Meat Chickens and Breeding Chickens. http://www.defra.gov.uk/animalh/welfare/farmed/meatchks/meatch kscode.pdf

154 RSPCA (1/9/2009). http://www.lookforthelogo.co.uk/ (1/9/2009)

155 RSPCA. Welfare Standards. http://www.rspca.org.uk/servlet/Satellite?pagename=RSPCA/RSPCA Redi- rect&pg=welfarestandards&marker=1&articleId=1121442811407

156 IMAGE CREDIT – This image is in the public domain because it is a mere mechanical scan or photocopy of a public domain original, or – from the available evidence – is so similar to such a scan or photocopy that no copyright protection can be expected to arise.http://commons.wikimedia.org/wiki/File:Taking_the_blubber_ off_a_whale_at_Nantucket,_by_Freeman,_J._(Josiah)_2.jpg

157 IMAGE CREDIT – Foodista. Reprodcued under a Creative Commons Attribution license (CC-BY): Attibution 2.0 Generic. http://www.flickr.com/photos/foodista/3276608628/

158 IMAGE CREDIT- Rasback. Permission is granted to copy, distribute and/or modify this document under the terms of the GNU Free Documentation License, Version 1.2 or any later version published by the Free Software Foundation; with no Invariant Sections, no Front-Cover Texts, and no Back-Cover Texts. http://commons.wikimedia.org/wiki/File:Portugiesischer-maulwurf- gebiss%26grabschaufeln.jpg

159 IMAGE CREDIT – Luca Galuzzi. This file is licensed under the Creative Commons Attribution ShareAlike 2.5 License. http://commons.wikimedia.org/wiki/File:Male_Lion_and_Cub_Chit wa_South_Africa_Luca_Galuzzi_2004.JPG

160 IMAGE CREDIT – Mahalie Stackpole. Reprodcued under a Creative Commons Attribution license (CC-BY): Attibution-Share Alike 2.0 Generic. http://www.flickr.com/photos/mahalie/292143657/

161 IMAGE CREDIT – Jeremy Keith. Reprodcued under a Creative Commons Attribution license (CC-BY): Attibution 2.0 Generic. http://www.flickr.com/photos/adactio/18845759/

162 IFAW (1/9/2009). http://www.stopwhaling.org/atf/cf/%7BC7E2199F-3FE3-487F- 90AC-8FF4FBB61804%7D/aboriginalsubsistencewhaling.PDF

163 Greenpeace (1/9/2009).
http://www.greenpeace.org/australia/resources/faqs/whales

164 WWF (1/9/2009). http://www.panda.org/?uNewsID=165541

165 WSPA. Troubled Waters. A review of the welfare implications of modern whaling activities.
http://www.wspa.org.uk/Images/080304_173035_TroubledWatersW haleReport_tcm9-2729.pdf

166 WSPA. Policy Statement. Available on request.

167 IFAW (1/9/2009).
http://www.stopwhaling.org/site/c.hhLTK0PDLqF/b.1389015/k.15/L earn_More_Whaling_Secrets_Youre_Not_Supposed_to_Know.htm

168 Charles Darwin Foundation. Pinta Giant Tortoise Fact Sheet.
http://www.darwinfoundation.org/english/pages/index.php (link removed, although I do have a copy).

169 Charles Darwin Foundation. Pinta Giant Tortoise Fact Sheet.
http://www.darwinfoundation.org/english/pages/index.php (link removed, although I do have a copy).

170 Charles Darwin Foundation. Pinta Giant Tortoise Fact Sheet.
http://www.darwinfoundation.org/english/pages/index.php (link removed, although I do have a copy).

171 Darwin, C. R. 1845. Journal of researches into the natural history and geology of the countries visited during the voyage of H.M.S. Beagle round the world, under the Command of Capt. Fitz Roy, R.N. 2d edition. Page 394 (available online at darwin-online.org.uk).

172 Charles Darwin Foundation. Pinta Giant Tortoise Fact Sheet.
http://www.darwinfoundation.org/english/pages/index.php (link removed, although I do have a copy).

173 Darwin, C. R. 1839. Narrative of the surveying voyages of His Majesty's Ships Adventure and Beagle between the years 1826 and 1836, describing their examination of the southern shores of South America, and the Beagle's circumnavigation of the globe. Journal and remarks. Page 459 (available online at darwin-online.org.uk).

174 Lonesome George is not alone among Galapagos tortoises. Current Biology (1/5/2007).
http://www.bio.mq.edu.au/molecularecology/pdfs/publications/CBt or.pdf

175 Charles Darwin Foundation. Pinta Giant Tortoise Fact Sheet.
http://www.darwinfoundation.org/english/pages/index.php (link removed, although I do have a copy).

[176] Charles Darwin Foundation. Pinta Giant Tortoise Fact Sheet. http://www.darwinfoundation.org/english/pages/index.php (link removed, although I do have a copy).

[177] Nigel Dougherty. Oblivion can wait for the Takahē. Veterinary Times (19/1/2009).

[178] George Orwell. Animal Farm (1945). Different context but the message still applies.

[179] Holger Kreft (1/9/2009). http://www-biology.ucsd.edu/news/article_051109.html

[180] Esmond Bradley. New Scientist Magazine (2003). http://www.newscientist.com/article/dn3376

[181] WWF (1/9/2009). http://www.panda.org/about_our_earth/teacher_resources/best_place_species/back_from_the_brink/saiga.cfm

[182] WWF (1/9/2009). http://www.panda.org/faq/response.cfm?hdnQuestionId=23520050111532

[183] WWF (1/9/2009). http://www.saigak.biodiversity.ru/eng/publications/wwf.html

[184] IUCN. African elephants and rhinos status survey and conservation action plan. http://www.rhinoresourcecenter.com/index.php?s=1&act=refs&CODE=ref_detail&id=1165235581

[185] IUCN. African elephants and rhinos status survey and conservation action plan. http://www.rhinoresourcecenter.com/index.php?s=1&act=refs&CODE=ref_detail&id=1165235581

[186] Sophie Stafford. BBC Wildlife Magazine (4/2009)

[187] Green10. 10 simple steps to help halt biodiversity loss by 2010. green10http://www.foeeurope.org/publications/2006/Green_10_biodiversity_May2006.pdf

[188] WWF. Magazine flyer.

[189] WWF. Magazine flyer.

[190] IFAW. Campaign email.

[191] WWF. WWF position on whaling and the IWC. http://assets.panda.org/downloads/wwf_position_iwc60_final.pdf

[192] Greenpeace (1/9/2009). http://www.greenpeace.org/australia/resources/faqs/whales

[193] IMAGE CREDIT – Ansgar Walk. This file is licensed under the Creative Commons Attribution ShareAlike 2.5 License. http://commons.wikimedia.org/wiki/File:Bowhead_Whale_2002-08-10.jpg

[194] IUCN (1/9/2009). http://www.iucnredlist.org/details/2474/0

[195] World Conservation Society (1/9/2009). http://www.bronxzoo.com/plan-your-trip/exhibits/tiger-mountain.aspx

[196] ZSL. http://static.zsl.org/files/tiger-conservation-41.pdf

[197] Heather Sohl (1/9/2009). http://www.wwf.org.uk/article_search_results.cfm?uNewsID=539

[198] Save the Tiger Fund. http://www.savethetigerfund.org/ (actual link removed. I have a copy).

[199] 'Global Tiger Patrol (1/9/2009). http://www.globaltigerpatrol.org/

[200] British Natural History Museum. From the Beginning Galley.

[201] Julie Roberts. Slug deterrents. BBC Wildlife Magazine (3/2009).

[202] BBC Wildlife Magazine (1/9/2009). http://info.bbcwildlifemagazine.com/about.php?jchk=1&nolog=1&jlnk=lsl0020

[203] The big question…What is evolution? BBC Wildlife Magazine (8/2008).

[204] Jeffrey McNeely (1/9/2009). Charles Darwin, a man for our time. http://www.iucn.org/es/noticias/news/?2650/Charles-Darwin-a-man-for-our-time

[205] Flash Recovery Of Ammonoids After Most Massive Extinction Of All. (16/9/2009). Timehttp://www.sciencedaily.com/releases/2009/09/090902122331.htm

[206] Mark Carwardine. BBC Wildlife Magazine (10/2006). http://www.markcarwardine.com/articles/bbc_wildlife/oct_2006.pdf

[207] Phylogeography and Genetic Ancestry of Tigers (Panthera tigris). PloS Biology (12/2004). http://www.plosbiology.org/article/info:doi/10.1371/journal.pbio.0020442

[208] The Giraffe is split into six seperate species. BBC Wildlife Magazine (4/2008).

[209] Multi-gene evidence for a new bottlenose dolphin species in southern Australia. Molecular Phylogenetics and Evolution (2008).

http://www.bio.mq.edu.au/molecularecology/pdfs/publications/new_sp_mpe08.pdf

[210] Cryptic diversity in vertebrates: molecular data doubles estimates of species diversity in a radiation of Australian lizards (Diplodactylus, Gekkota). Proceedings of the Royal Society: B (4/3/2009). http://rspb.royalsocietypublishing.org/content/early/2009/02/27/rspb.2008.1881.abstract

[211] 'RSPB. Conservation Science in the RSPB. http://www.rspb.org.uk/Images/ConSciReport2001_tcm9-133174.pdf

[212] RSPB. Conservation Science in the RSPB. http://www.rspb.org.uk/Images/ConSciReport2001_tcm9-133174.pdf

[213] 'Biochemical studies revealed no significant differentiation between Scottish, parrot and common crossbills in their mitochondrial DNA and microsatellites.'

[214] RSPB. Conservation Science in the RSPB. http://www.rspb.org.uk/Images/ConSciReport2001_tcm9-133174.pdf

[215] Ron Summers (1/9/2009). http://news.bbc.co.uk/1/hi/scotland/highlands_and_islands/4793863.stm

[216] Congenital bone deformities and the inbred wolves (Canis lupis) of Isla Royale. Biological Conservation (5/2009). http://www.isleroyalewolf.org/overview/overview/rescue

[217] The Wolves and Moose of Isla Royale (1/9/2009). http://www.isleroyalewolf.org/overview/overview/rescue

[218] The Wolves and Moose of Isla Royale (1/9/2009). http://www.isleroyalewolf.org/overview/overview/at_a_glance.html

[219] The Wolves and Moose of Isla Royale (1/9/2009). http://www.isleroyalewolf.org/overview/overview/rescue

[220] Congenital bone deformities and the inbred wolves (Canis lupis) of Isla Royale. Biological Conservation (5/2009). http://www.isleroyalewolf.org/overview/overview/rescue

[221] Kris Hundertmark (1/9/2009). http://www.uaf.edu/news/a_news/20090618143756.html

[222] Richard Dawkins. The God Delusion (2006).

[223] Animals that seem identical may be completely different species (1/9/2009).

http://www.science.gu.se/english/News/News_detail/_Animals_that
_seem_identical_may_be_completely_different_species.cid878699
(2/9/2009).

[224] Yale University (1/9/2009).
http://opa.yale.edu/news/article.aspx?id=6049

[225] The Quagga Project (1/9/2009). http://www.quaggaproject.org/

[226] The Quagga Project (1/9/2009).
http://media1.mweb.co.za/quaggaproject/quagga-faq.htm

[227] The Quagga Project (1/9/2009).
http://media1.mweb.co.za/quaggaproject/quagga-species.htm

[228] The Quagga Project (1/9/2009).
http://media1.mweb.co.za/quaggaproject/quagga-extinction-
quagga.htm

[229] The Quagga Project (1/9/2009).
http://media1.mweb.co.za/quaggaproject/quagga-criticism.htm

[230] IMAGE CREDIT - This image (or other media file) is in the
public domain because its copyright has expired.
http://commons.wikimedia.org/wiki/File:Quagga_photo.jpg

[231] Saving the Barbary Lion. BBC Wildlife Magazine (4/2009).

[232] IAEA. Chernobyl's Legacy.
http://www.iaea.org/Publications/Booklets/Chernobyl/chernobyl.pdf

[233] Greenpeace. 'Page Not Found' website response.

[234] IAEA. Chernobyl's Legacy.
http://www.iaea.org/Publications/Booklets/Chernobyl/chernobyl.pdf

[235] IAEA. Environmental Consequences of the Chernobyl Accident
and their Remediation. http://www-
pub.iaea.org/MTCD/publications/PDF/Pub1239_web.pdf

[236] IAEA. Environmental Consequences of the Chernobyl Accident
and their Remediation. http://www-
pub.iaea.org/MTCD/publications/PDF/Pub1239_web.pdf

[237] IAEA. Chernobyl's Legacy.
http://www.iaea.org/Publications/Booklets/Chernobyl/chernobyl.pdf

[238] IAEA. Health Effects of the Chernobyl Accident.
http://www.who.int/ionizing_radiation/chernobyl/who_chernobyl_r
eport_2006.pdf

[239] James Lovelock. Gaia: A New Look at Life on Earth (1979
Paperback), page 1.

[240] James Lovelock. Gaia: A New Look at Life on Earth (1979 Paperback), page 142.

[241] James Lovelock. Gaia: A New Look at Life on Earth (1979 Paperback), page 37.

[242] James Lovelock. The Revenge of Gaia (2006 Paperback), page 190.

[243] James Lovelock. Gaia: A New Look at Life on Earth (1979 Paperback), page 117.

[244] James Lovelock. The Vanishing Face of Gaia (2009 Hardback), page 103.

[245] James Lovelock. The Vanishing Face of Gaia (2009 Hardback), page 49.

[246] James Lovelock. The Vanishing Face of Gaia (2009 Hardback), page 151.

[247] James Lovelock. Gaia Series.

[248] James Lovelock. The Vanishing Face of Gaia (2009 Hardback), page 12.

[249] James Lovelock. The Vanishing Face of Gaia (2009 Hardback), page 57.

[250] James Lovelock. The Vanishing Face of Gaia (2009 Hardback), page 12.

[251] James Lovelock. The Revenge of Gaia (2006 Paperback), page 76.

[252] James Lovelock. The Revenge of Gaia (2006 Paperback), page 33.

[253] James Lovelock. Gaia: A New Look at Life on Earth (1979 Paperback), page 100.

[254] James Lovelock. The Vanishing Face of Gaia (2009 Hardback), page 57.

[255] James Lovelock. The Revenge of Gaia (2006 Paperback), page 76.

[256] James Lovelock. The Vanishing Face of Gaia (2009 Hardback), page 155.

[257] James Lovelock. Gaia: A New Look at Life on Earth (1979 Paperback), page 28.

[258] Sedimentary challenge to Snowball Earth. Nature Geoscience (12/2008).
http://www.nature.com/ngeo/journal/v1/n12/abs/ngeo355.html

[259] British Natural History Museum. From the Beginning Gallery.

260 WWF. A Roadmap for a Living Planet.
http://assets.panda.org/downloads/roadmap_sign_off_fin.pdf

261 WWF (2/9/2009). http://www.worldwildlife.org/who/Vision/

262 WWF. A Roadmap for a Living Planet.

263 WWF (2/9/2009). http://www.worldwildlife.org/who/Vision/

264 Wildlife Conservation Society (2/9/2009).
http://www.wcs.org/about-us.aspx

265 Wildlife Conservation Society (2/9/2009).
'http://archive.wcs.org/globalconservation/.'

266 Conservation International (2/9/2009).
http://blog.conservation.org/about/

267 Kofi Anan (2/9/2009).
http://www.unis.unvienna.org/unis/pressrels/2001/sgsm7818.html

268 An Open Letter to the American People (2004) (2/9/2009).
http://www.mellicant.com/scientists/nobelletter.php

269 Mark Carwardine. Wild Thoughts. BBC Wildlife Magazine
(7/2006).

270 Sophie Stafford. Editor's Letter. BBC Wildlife Magazine
(10/2006).

271 British Natural History Museum. Ecology Gallery.

272 British Natural History Museum. Restless Surface Gallery.

273 British Natural History Museum. Restless Surface Gallery.

274 British Natural History Museum. Restless Surface Gallery.

275 British Natural History Museum. Restless Surface Gallery.

276 British Natural History Museum. Restless Surface Gallery.

277 British Natural History Museum. Restless Surface Gallery.

278 British Natural History Museum. Restless Surface Gallery.

279 British Natural History Museum. From the Beginning Gallery.

280 British Natural History Museum. Restless Surface Gallery. Earth
Toad and Tomorrow.

281 Greenpeace (2/9/2009). http://www.greenpeace.org/international/

282 RSPB (2/9/2009). http://www.rspb.org.uk/

283 NRDC (2/9/2009). http://www.nrdc.org/naturesvoice/default.asp

284 James Lovelock. The Vanishing Face of Gaia (2009 Hardback),
page 6.

[285] British Natural History Museum (2/9/2009).
http://www.nhm.ac.uk/nature-online/insite/discovering-
understanding/F31.html

[286] WWF.
http://assets.panda.org/downloads/resolving_ddt_summ_english.pdf

[287] James Lovelock. The Revenge of Gaia (2006), page 116-117.

[288] IUCN. Transition to Sustainability. Towards a Humane and
Diverse World. http://data.iucn.org/dbtw-wpd/edocs/2008-017.pdf

[289] James Lovelock. The Revenge of Gaia (2006 Paperback), page 9.

[290] IPCC (2003). Climate Change 2001: The Scientific Basis, page
201. http://www.grida.no/publications/other/ipcc_tar/

[291] IPCC (2003). Climate Change 2001: The Scientific Basis, page
201. http://www.grida.no/publications/other/ipcc_tar/

[292] WWF (2/9/2009).
http://www.wwf.org.uk/what_we_do/tackling_climate_change/clima
te_change_explained/index.cfm

[293] Al Gore - An Inconvenient Truth (2006).

[294] George Monbiot. Heat: How we can stop the planet burning
(2007).

[295] Advanced Greenhouses (2/9/2009).
http://www.advancegreenhouses.com/greenhouse_co2_generators_f
rom_a.htm

[296] Scott Denning (13/9/2009).
http://www.sciencedaily.com/videos/2007/0603-
can_carbon_dioxide_be_a_good_thing.htm

[297] IPCC (2003). Climate Change 2001: The Scientific Basis, page
202. http://www.grida.no/publications/other/ipcc_tar/

[298] IPCC (2003). Climate Change 2001: The Scientific Basis, page
202. http://www.grida.no/publications/other/ipcc_tar/

[299] WWF (2/9/2009).
http://www.wwf.org.uk/what_we_do/tackling_climate_change/clima
te_change_explained/index.cfm

[300] British Natural History Museum. From the Beginning Gallery.

[301] IPCC (2007). Climate Change 2007: The Physical Science Basis,
page 465. http://www.ipcc.ch/pdf/assessment-report/ar4/wg1/ar4-
wg1-chapter6.pdf

[302] IPCC (2007). Climate Change 2007: The Physical Science Basis, page 465. http://www.ipcc.ch/pdf/assessment-report/ar4/wg1/ar4-wg1-chapter6.pdf

[303] IPCC (2007). Climate Change 2007: The Physical Science Basis, page 442. http://www.ipcc.ch/pdf/assessment-report/ar4/wg1/ar4-wg1-chapter6.pdf

[304] IPCC (2007). Climate Change 2007: The Physical Science Basis, page SM.6-3. http://www.ipcc.ch/pdf/assessment-report/ar4/wg1/ar4-wg1-chapter6-supp-material.pdf

[305] National Geographic Magazine (2/9/2009). http://ngm.nationalgeographic.com/ngm/0505/resources_geo.html.

[306] National Geographic Magazine (2/9/2009). http://ngm.nationalgeographic.com/ngm/0505/resources_geo.html.

[307] Darwin Centre for Biogeology (2/9/2009). http://www3.bio.uu.nl/palaeo/Azolla/ecophysiological.htm

[308] Cardiff University (13/9/2009). http://www.cardiff.ac.uk/news/mediacentre/mediareleases/Sept09/new-co2-data-helps-unlock-the-secrets-of-antarctic-formation.html

[309] IPCC. Climate Change (2007) The Physical Science Basis. http://ipcc-wg1.ucar.edu/wg1/FAQ/wg1_faq-6.1.html

[310] IPCC (2007). Climate Change 2007: The Physical Science Basis, page 440. http://www.ipcc.ch/pdf/assessment-report/ar4/wg1/ar4-wg1-chapter6.pdf

[311] James Lovelock. The Revenge of Gaia (2006 Paperback), page 80.

[312] James Lovelock. The Revenge of Gaia (2006 Paperback), page 80.

[313] James Lovelock. The Vanishing Face of Gaia (2009), page 101.

[314] IPCC (2007). Climate Change 2007: The Physical Science Basis, page 440. http://www.ipcc.ch/pdf/assessment-report/ar4/wg1/ar4-wg1-chapter6.pdf

[315] James Lovelock. The Vanishing Face of Gaia (2009 Hardback), page 44

[316] WWF. Magazine flyer.

[317] WWF. Magazine flyer.

[318] WWF. Polar Bear. http://www.worldwildlife.org/species/finder/polarbear/WWFBinaryitem5832.pdf

[319] WWF. Magazine flyer.

[320] WWF. Polar Bear.
http://www.worldwildlife.org/species/finder/polarbear/WWFBinaryi tem5832.pdf

[321] Greenpeace (2/9/2009).
http://www.greenpeace.org/international/campaigns/oceans/pollutio n/creeping-dead-zones

[322] NASA (2/9/2009).
http://www.nasa.gov/vision/earth/environment/dead_zone.html

[323] NSF (2/9/2009).
http://www.nsf.gov/news/special_reports/jellyfish/textonly/locations _gulfmexico.jsp

[324] University of Copenhagen (2/9/2009). Dramatic Expansion of Dead Zones in the Ocean.
http://www.nbi.ku.dk/english/news/dead_zones_in_the_oceans/

[325] Brown University (2/9/2009). Brown Scientist Finds Coastal Dead Zones May Benefit Some Species.
http://news.brown.edu/pressreleases/2008/10/deadzones

[326] RSPB (2/9/2009).
http://www.rspb.org.uk/ourwork/conservation/managingreserves/ha bitats/reedbeds/reedbedcreation.asp

[327] British Dragonfly Society (2/9/2009).
http://www.dragonflysoc.org.uk/mffispumfull.htm

[328] British Dragonfly Society (2/9/2009).
http://www.dragonflysoc.org.uk/mffispumfull.htm.

[329] Herpetological Conservation Trust (2/9/2009).
http://www.herpconstrust.org.uk/hab-management.htm

[330] Herpetological Conservation Trust (2/9/2009).
http://www.herpconstrust.org.uk/hab-management.htm

[331] Herpetological Conservation Trust (2/9/2009).
http://www.herpconstrust.org.uk/hab-management.htm

[332] 'Herpetological Conservation Trust (2/9/2009).
http://www.herpconstrust.org.uk/hab-management.htm

[333] Responsibletravel.com (2/9/2009).
http://www.responsibletravel.com/trip/Trip100211.htm

[334] Paul Evans.
http://www.guardian.co.uk/environment/2009/may/29/conservation -reintroductions-nature

[335] WWF (2/9/2009).
http://www.panda.org/about_our_earth/teacher_resources/project_ideas/balanced_ecosystem/

[336] James Lovelock. Gaia: A New Look at Life on Earth (1979 Paperback), page 140.

[337] Carl Gustaf Lundin (2/9/2009).
http://www.iucn.org/knowledge/news/focus/

[338] WWF (2/9/2009). http://www.panda.org/who_we_are/

[339] Flora and Fauna International. Magazine flyer.

[340] Wildlife Conservation Society (2/9/2009).
http://www.wcs.org/press/press-releases/high-rates-in-wildlife-should-concern-humans.aspx

[341] Friends of The Earth (2/9/2009).
http://www.foeeurope.org/about/english.htm

[342] Conservation International (2/9/2009).
http://blog.conservation.org/about/

[343] IUCN. A 2020 Vision for IUCN.
http://cmsdata.iucn.org/downloads/2020_vision_for_iucn_en.pdf

[344] RSPCA (2/9/2009).
http://www.rspca.org.uk/servlet/Satellite?pagename=RSPCA/RSPCARedirect&pg=about_the_rspca

[345] British NaturL History Museum. Begging box.

[346] David Klein. The introduction, Increase, and Crash of Reindeer on St. Matthew Island. http://dieoff.org/page80.htm

[347] WWF (2/9/2009).
http://www.panda.org/what_we_do/endangered_species/tigers/tigers_ecology_habitat/

[348] Data from the United Nations Population Division. The World at Six Billion.
http://www.un.org/esa/population/publications/sixbillion/sixbillion.htm

[349] United Nations Population Division. World Population Prospects: The 2008 Revision.
http://www.un.org/esa/population/publications/wpp2008/pressrelease.pdf

[350] National Geographic Magazine (2/9/2009).
http://ngm.nationalgeographic.com/2009/06/cheap-food/bourne-text

[351] National Geographic Magazine (2/9/2009).
http://ngm.nationalgeographic.com/2009/06/cheap-food/bourne-text

[352] National Geographic Magazine (2/9/2009).
http://ngm.nationalgeographic.com/2009/06/cheap-food/bourne-text

[353] National Geographic Magazine (2/9/2009).
http://ngm.nationalgeographic.com/2009/06/cheap-food/bourne-text

[354] National Geographic Magazine (2/9/2009).
http://ngm.nationalgeographic.com/2009/06/cheap-food/bourne-text

[355] National Geographic Magazine (2/9/2009).
http://ngm.nationalgeographic.com/2009/06/cheap-food/bourne-text

[356] Richard Dawkins. The Selfish Gene (1976).

[357] James Lovelock. The Vanishing Face of Gaia (2009 Hardback), page 156.

[358] James Lovelock. The Vanishing Face of Gaia (2009 Hardback), page 150.

[359] Guardian. World is facing a natural resources crisis worse than financial crunch.
http://www.guardian.co.uk/environment/2008/oct/29/climatechange-endangeredhabitats

[360] Guardian. Heart disease gene affects one in 100, say scientists.
http://www.guardian.co.uk/science/2009/jan/18/heart-disease-gene

[361] British Natural History Museum. Ecology Gallery.

[362] British Natural History Museum. Ecology Gallery.

[363] Theos. Faith and Darwin: Faith, Harmony, or Confusion.
http://campaigndirector.moodia.com/Client/Theos/Files/FaithandDarwin.pdf

[364] James Lovelock. The Vanishing Face of Gaia (2009 Hardback), page 6.

[365] James Lovelock. The Vanishing Face of Gaia (2009 Hardback), page 21.

[366] James Lovelock. Guardian Interview: James Lovelock on how to save Gaia (2009).

http://www.guardian.co.uk/science/video/2009/apr/22/james-lovelock-gaia-space-biochar

[367] James Lovelock. The Vanishing Face of Gaia (2009 Hardback), page 63.

[368] James Lovelock. The Vanishing Face of Gaia (2009 Hardback), page 52.

[369] James Lovelock. The Vanishing Face of Gaia (2009 Hardback), page 156.

[370] James Lovelock. The Vanishing Face of Gaia (2009 Hardback), page 123.

[371] Richard Dawkins. The Selfish Gene (1976).

[372] 'In most case, the demise of a place or species can be directly attributable to one of humankind's many sins against the planet.' Source: WWF.

[373] Fauna and Flora International. Magazine flyer.

[374] Fauna and Flora International. Magazine flyer.

[375] Fauna and Flora International. Magazine flyer.

[376] Richard Dawkins. The God Delusion (2006).

[377] WWF. Magazine flyer.

[378] WWF. Magazine flyer.

[379] Mark Carwardine. Wild Thoughts. BBC Wildlife Magazine (10/2006).
http://www.markcarwardine.com/articles/bbc_wildlife/oct_2006.pdf

[380] The Orangutan Foundation (7/9/2009).
http://www.orangutan.org.uk/

[381] The Orangutan Foundation (7/7/2009).
http://www.orangutan.org.uk/about-orangutans/species-information

[382] WWF. Magazine flyer.

[383] Greenpeace (2/9/2009).
http://www.greenpeace.org/international/supportus

[384] British Natural History Museum. Begging box.

[385] Wildlife Conservation Society (2/9/2009).
http://archive.wcs.org/getinvolved/donations.html

[386] Red Squirrel Survival Trust (2/9/2009).
http://www.rsst.org.uk/index.php?option=com_content&view=article&id=47&Itemid=82

[387] IFAW (2/9/2009).
http://www.stopwhaling.org/c.foJNIZOyEnH/b.5232437/k.12A7/Ur
ge_Your_Representative_to_CoSponsor_the_Whale_Conservation_
and_Protection_Act_of_2009/siteapps/advocacy/ActionItem.aspx

[388] IFAW. Campaign email.

[389] Rebecca Aldworth. Campaign email.

[390] Rebecca Aldworth. Campaign email.

[391] Rebecca Aldworth. Campaign email.

[392] IFAW (link removed).'

[393] WWF (2/9/2009).
http://www.panda.org/what_we_do/endangered_species/rhinoceros
/african_rhinos/rhino_conservation/

[394] WWF (2/9/2009).
http://www.panda.org/wwf_news/press_releases/?uNewsID=169861

[395] WWF (2/9/2009).
http://www.panda.org/what_we_do/endangered_species/rhinoceros
/african_rhinos/rhino_conservation/

[396] WWF (2/9/2009).
http://www.panda.org/what_we_do/endangered_species/rhinoceros
/african_rhinos/rhino_conservation/

[397] WWF. Campaign email.

[398] WWF (2/9/2009).
http://www.panda.org/what_we_do/endangered_species/giant_pand
a/pandasuccess/

[399] WWF (2/9/2009).
http://www.panda.org/what_we_do/endangered_species/giant_pand
a/pandasuccess/

[400] 'WWF (2/9/2009).
http://www.panda.org/what_we_do/endangered_species/giant_pand
a/pandasuccess/

[401] Harvard Business School (2/9/2009).
http://hbswk.hbs.edu/item/5797.html

[402] WWF (2/9/2009). http://www.panda.org/who_we_are/

[403] Harvard Business School (2/9/2009).
http://hbswk.hbs.edu/item/5797.html

[404] UNEP. UNEP Year Book 2009.
http://www.unep.org/geo/yearbook/yb2009/PDF/5-
Resource_Efficiency_UNEP_YearBook_09_low.pdf

[405] WWF. Magazine flyer.

[406] WWF. Magazine flyer.

[407] UNEP. UNEP Year Book 2009.
http://www.unep.org/geo/yearbook/yb2009/PDF/3-
Climate_Change_%20UNEP_YearBook_09_low.pdf

[408] WWF. Magazine flyer.

[409] WWF. Polar Bear.
http://www.worldwildlife.org/species/finder/polarbear/WWFBinaryi
tem5832.pdf

[410] WWF. Magazine flyer.

[411] IUCN (2/9/2009). http://www.iucnredlist.org/details/22823/0

[412] Arizona Game and Fish Department Jaguar Conservation team
(2/9/2009). http://www.azgfd.gov/w_c/es/jaguar_management.shtml

[413] Arizona Game and Fish Department Jaguar Conservation team
(2/9/2009). http://www.azgfd.gov/w_c/es/jaguar_management.shtml

[414] Arizona Game and Fish Department Jaguar Conservation team
(2/9/2009). http://www.azgfd.gov/w_c/es/jaguar_management.shtml

[415] Arizona Game and Fish Department Jaguar Conservation team
(2/9/2009). http://www.azgfd.net/wildlife/conservation-
news/arizona-game-and-fish-collars-first-wild-jaguar-in-united-
states/2009/02/20/

[416] Arizona Game and Fish Department Jaguar Conservation team
(2/9/2009). http://www.azgfd.gov/w_c/jaguar/TimeLine.shtml

[417] Arizona Game and Fish Department Jaguar Conservation team
(2/9/2009). http://www.azgfd.gov/w_c/jaguar/TimeLine.shtml

[418] Arizona Game and Fish Department Jaguar Conservation team
(2/9/2009). http://www.azgfd.gov/w_c/jaguar/JaguarFAQ.shtml

[419] Arizona Game and Fish Department Jaguar Conservation team
(2/9/2009). http://www.azgfd.gov/w_c/jaguar/Message.shtml

[420] Steve Spangle.
http://www.fws.gov/southwest/es/arizona/Documents/SpeciesDocs/J
aguar/090302_Jaguarrecapture_nr_v5.pdf

[421] Arizona Game and Fish Department Jaguar Conservation team
(2/9/2009). http://www.azgfd.gov/w_c/jaguar/JaguarFAQ.shtml.

[422] Emil McCain.
http://www.azgfd.gov/w_c/jaguar/documents/Jaguar.Mythoughtson
MachoB.E_McCain.20090303.pdf

[423] Emil McCain.
http://www.azgfd.gov/w_c/jaguar/documents/Jaguar.Mythoughtson
MachoB.E_McCain.20090303.pdf

[424] Emil McCain.
http://www.azgfd.gov/w_c/jaguar/documents/Jaguar.Mythoughtson
MachoB.E_McCain.20090303.pdf

[425] and Emil McCain.
http://www.azgfd.gov/w_c/jaguar/documents/Jaguar.Mythoughtson
MachoB.E_McCain.20090303.pdf

[426] Emil McCain.
http://www.azgfd.gov/w_c/jaguar/documents/Jaguar.Mythoughtson
MachoB.E_McCain.20090303.pdf

[427] Larry Voyles. http://www.azgfd.gov/w_c/jaguar/Message.shtml

[428] Arizona Republic newspaper.
http://www.azgfd.gov/w_c/jaguar/ArizonaRepublic.shtml

[429] Arizona Game and Fish Department.
http://www.azgfd.gov/h_f/hunting.shtml

[430] Arizona Game and Fish Department.
http://www.azgfd.gov/h_f/biggame_species.shtml

[431] Arizona Game and Fish Department.
http://www.azgfd.gov/h_f/game_lion.shtml

[432] Arizona Game and Fish Department Jaguar Conservation team
(2/9/2009). http://www.azgfd.net/wildlife/conservation-
news/international-conservation-program-brings-endangered-
jaguar-to-arizona-only-wild-born-jaguar-in-an-american-
zoo/2008/11/24/

[433] Arizona Game and Fish Department Jaguar Conservation team.
http://www.azgfd.gov/pdfs/w_c/jaguar/jaguar_brochure.pdf

[434] Arizona Game and Fish Department Jaguar Conservation team
(2/9/2009).http://www.azgfd.gov/w_c/es/jaguar_management.shtml

[435] Greenpeace (3/9/2009).
http://www.greenpeace.org/international/news/10years-amazon-
rainforest060509'

[436] FAO. FAOSTAT. http://faostat.fao.org/default.aspx

[437] WWF (2/9/2009).
http://www.panda.org/what_we_do/footprint/climate_carbon_ener
gy/forest_climate/

[438] Greenpeace (2/9/2009).
http://www.greenpeace.org/international/about/reports

[439] IUCN. A 2020 Vision for IUCN.
http://cmsdata.iucn.org/downloads/2020_vision_for_iucn_en.pdf

[440] IUCN. A 2020 Vision for IUCN.
http://cmsdata.iucn.org/downloads/2020_vision_for_iucn_en.pdf

[441] IUCN. Transition to Sustainability: Towards a Humane and Diverse World. http://data.iucn.org/dbtw-wpd/edocs/2008-017.pdf

[442] IUCN (2/9/2009). http://www.iucn.org/about/

[443] IUCN. Transition to Sustainability: Towards a Humane and Diverse World. http://data.iucn.org/dbtw-wpd/edocs/2008-017.pdf

[444] IUCN. Transition to Sustainability: Towards a Humane and Diverse World. http://data.iucn.org/dbtw-wpd/edocs/2008-017.pdf

[445] David Attenborough.
http://www.optimumpopulation.org/releases/opt.release13Apr09.htm

[446] James Lovelock. The Vanishing Face of Gaia (2009), page 3.

[447] WWF. Living Planet Report 2008.
http://assets.panda.org/downloads/living_planet_report_2008.pdf

[448] WWF. Living Planet Report 2008.
http://assets.panda.org/downloads/living_planet_report_2008.pdf

[449] IUCN. Transition to Sustainability: Towards a Humane and Diverse World. http://data.iucn.org/dbtw-wpd/edocs/2008-017.pdf